T0258491

Encyclopedia of Gas Chromatography: Selected Applications and New Techniques

Volume II

Encyclopedia of Gas Chromatography: Selected Applications and New Techniques Volume II

Edited by **Carol Evans**

NY RESEARCH
P R E S S

New York

Published by NY Research Press,
23 West, 55th Street, Suite 816,
New York, NY 10019, USA
www.nyresearchpress.com

Encyclopedia of Gas Chromatography: Selected Applications and New Techniques
Volume II
Edited by Carol Evans

International Standard Book Number: 978-1-63238-129-3 (Hardback)

Contents

Preface

This book aims to highlight the current researches and provides a platform to further the scope of innovations in this area. This book is a product of the combined efforts of many researchers and scientists, after going through thorough studies and analysis from different parts of the world. The objective of this book is to provide the readers with the latest information of the field.

This book is a collection of studies on selected applications and new techniques in gas chromatography. Diverse uses varying from basic biological applications to industrial uses have been overviewed in this book. Analysis of multidimensional gas chromatography – time of flight mass spectrometry of PAH in smog chamber studies and in smog samples, inverse gas chromatography in characterization of composites interaction has been presented in detail. The various topics have been addressed by masters from diverse spheres and supplemented by easy-to-understand figures and tables. It has been specially purposed for chemists and other practitioners working in the domains related to gas chromatography. This book is a valued resource for both budding and expert chromatographers.

I would like to express my sincere thanks to the authors for their dedicated efforts in the completion of this book. I acknowledge the efforts of the publisher for providing constant support. Lastly, I would like to thank my family for their support in all academic endeavors.

Editor

Part 1

Selected Application of
Gas Chromatography in Life Sciences

Application of Pyrolysis-Gas Chromatography/Mass Spectrometry to the Analysis of Lacquer Film

Rong Lu, Takayuki Honda and Tetsuo Miyakoshi
Department of Applied Chemistry,
School of Science and Technology, Meiji University,
1-1-1 Higashi-mita, Tama-ku, Kawasaki-shi,
Japan

1. Introduction

1.1 Characteristics of lacquer

Oriental lacquer is a reproducible natural product that has been used for thousands of years in Asia. No organic solvent evaporates during the drying process, only water. Because of the self-drying system, natural lacquer is an eco-friendly product that is expected to be useful in the future as a coating material.

Lacquer trees are members of the family Anacardiaceae, which includes more than 73 genera worldwide and approximately 600 species including mango (*Mangifera indica*) and cashew (*Anacardium occidentale*), but most of them grow in the subtropical region of Southeast Asia [1-3]. In general, only a few kinds of lacquer trees that grow in the evergreen forests of East Asia are able to produce lacquer sap.

Lacquer has been used in Asian countries for thousands of years as a durable and beautiful coating material [4-7]. Cultural treasures coated with lacquer have maintained their beautiful surfaces without loss of their original beauty for more than 9000 years [8-10].

Lacquer sap for the manufacture of lacquerware is collected not only in China and Japan but also in Southeast Asia, as shown in Figure 1-1. In Japan, about 10 years after cultivating a lacquer tree, cuts are made in the tree trunk. A white, milky resin seeps out of the wounds, and this sap is collected in a keg. After the lacquer sap has been collected, the tree is left untouched for 3 or 4 days before cuts are again made in the tree to collect more sap. Laccol (MW=348) is the main component of *Rhus succedanea* trees, which grow in Vietnam and Chinese Taiwan, and thitsiol (MW=348) is the main component of *Melanorrhoea (Gluta) usitata* trees, which grow in Thailand and Myanmar [11-13]. All these saps are used as a surface coating for wood, porcelain, and metal wares in Asia.

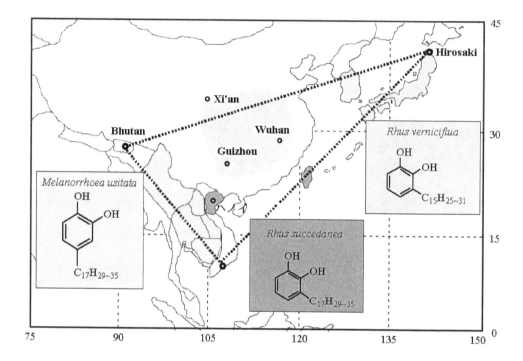

Fig. 1-1. Asian origin and type of lacquer sap

1.2 Components of lacquer

Lacquer is the sap obtained by tapping lac trees, specifically *Rhus vernicifera* (in Japan, China, and Korea), *Rhus succedanea* (in Vietnam and Taiwan), and *Melanorrhoea (Gluta) usitata* (in Thailand and Myanmar). The sap from *Rhus vernicifera* comprises a latex material composed of phenol derivatives (60-65%), water (20-25%), plant gum including saccharides (5-7%), water-insoluble glycoproteins (2-5%), and the laccase enzyme (0.1%) [11-13].

Under the microscope, this milky liquid appears to be a collection of water particles in an emulsion. Small particles of water are dispersed in the oily substance called urushiol. Unlike milk, in which fat is dispersed in water, water is dispersed in the urushiol oil in lacquer. The emulsion droplets contain the dissolved laccase enzyme and polysaccharides. It is believed that nitrogenous substances play an important role in forming a stable emulsion of the aqueous component and urushiol.

The composition of lacquer sap has been investigated using gel permeation chromatography (GPC), high performance liquid chromatography (HPLC), mass spectrometry (MS), and nuclear magnetic resonance (NMR), including two-dimensional NMR spectroscopy. Urushiol is a mixture of 3-substituted catechol derivatives having carbon chains with 15 carbons with 0-3 double bonds. The triene side chain of urushiol makes up 60-70% of urushiol [14]. All of these components have been separated and purified. Figure 1-2 shows photographs of urushiol, polysaccharides, glycoprotein, laccase, and stellacyanin obtained from Chinese lacquer sap isolated according to previously reported methods. Procedures utilized for the separation and purification of lacquer components from *Rhus vernicifera* lacquer sap are shown in Figure 1-3 [15].

Fig. 1-2. Photos of lacquer components

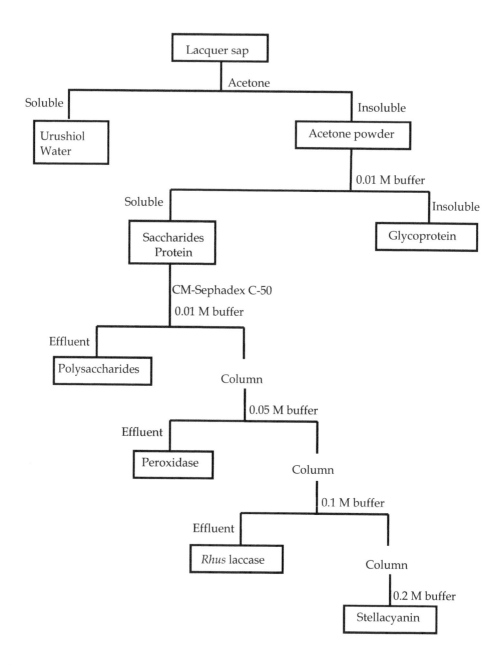

Fig. 1-3. Separation of lacquer sap

1.3 Drying mechanism

The drying mechanism of lacquer sap catalyzed by laccase was first explicated by Kumanotani [16], and the explanation was supplemented and developed by our later studies [17]. The catechol ring of urushiol is first oxidized by laccase to form dimers, trimers, and oligomers, and after the urushiol monomer concentration has decreased to less than 30%, a bridge-construction reaction due to auto-oxidation of the unsaturated side chain occurs. Figure 1-4 shows the laccase-catalyzed oxidation coupling of urushiol [16-17].

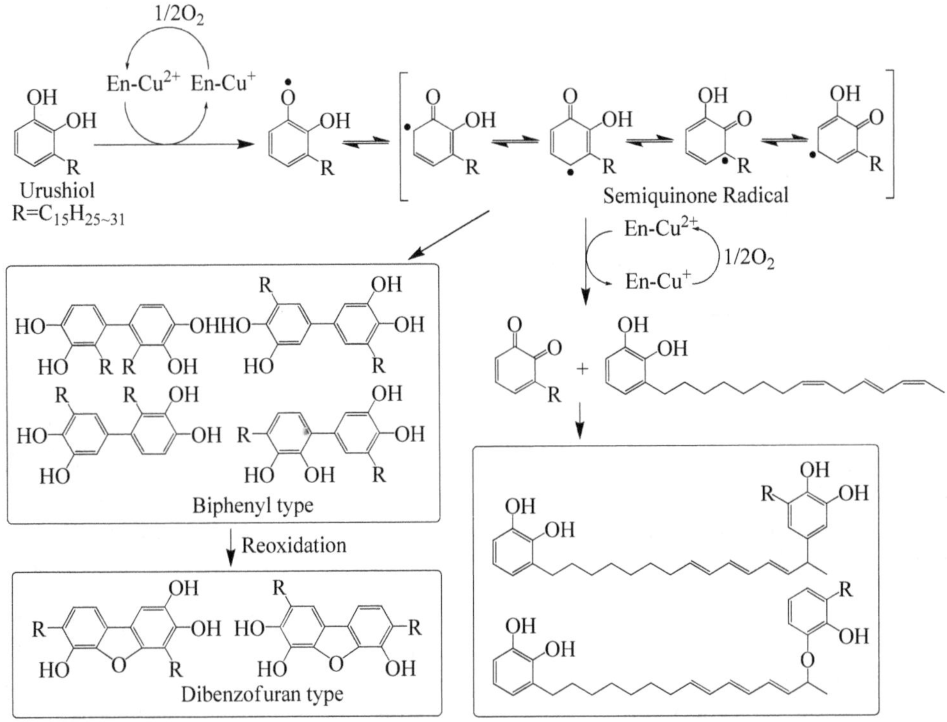

Fig. 1-4. Polymerization mechanism of urushiol catalyzed by laccase

The dimerization of urushiol proceeds through laccase-catalyzed nuclear-nuclear (C-C) coupling. Its detailed mechanism is as follows. During the first step of the dimerization, En-Cu^{2+} oxidizes urushiol to give the semiquinone radical and En-Cu^+. During the sap cooking process, the presence of plant gum cause significant foaming, which mixes air and sap during the treatment as well as accelerating the oxidation of Cu^+ to Cu^{2+} in the laccase. The reformed En-Cu^{2+} takes part in the repeated oxidation of urushiol. The semiquinone radical formed undergoes a C-C coupling reaction to produce biphenyl dimers, and the urushiol quinone formed through the disproportionation reaction undergoes hydrogen abstraction from the triene side chain of urushiol to give a nucleus-side chain C-C coupling dimer. The polymerization of urushiol may proceed through these types of couplings. However, many and various kinds of reactions may occur in the film-making process. Both enzymatic reaction and auto-oxidation (Figure 1-5) are repeated to form a durable network polymer.

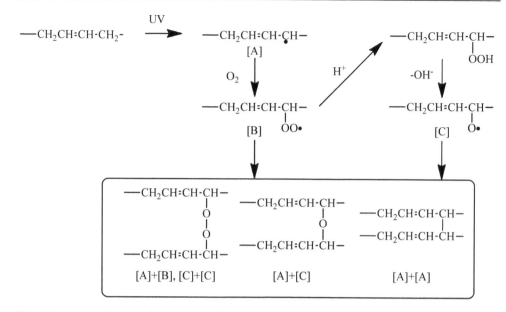

Fig. 1-5. Auto-oxidation of unsaturated lacquer side chain

In order to reveal the enzymatic polymerization mechanism of thitsiol, we reinvestigated the dimer structures produced in the initial stage of enzymatic polymerization by using modern techniques including 1H-1H and 1H-13C correlation two-dimensional NMR measurements [18-19]. The structures of thitsiol dimers from *Melanorrhoea (Gluta) usitata* were characterized by means of high-resolution NMR spectroscopy involving two-dimensional NMR measurements using field gradient DQF-COSY, HMQC, and HMBC. Almost all proton and carbon absorptions were assigned. The results showed that the main products of thitsiol dimers catalyzed by laccase are 1,1',2,2'-tetrahydroxy-3,3'-dialkenyl-5,5'-biphenyl, 1,1',2,2'-tetrahydroxy-3,3'-dialkenyl-6,5'-biphenyl, and 1,1',2,2'-tetrahydroxy-3,4'-dialkenyl-5,5'-biphenyl. The thitsiol dimers are almost all due to nuclear-nuclear (C-C) coupling, which differs from the nuclear-side chain (C-O-C) coupling of urushiol as shown in Figure 1-4 [20].

2. Py-GC/MS

Pyrolysis gas chromatography/mass spectrometry (Py-GC/MS) is a comparatively old analytical method (Figure 2-1). It is one of the techniques developed to analyze polymers. It has problems with reproducibility and long analysis times and it was not used very much, though pyrolysis began to be used to analyze natural fibers and synthetic fibers. The technique for combining this pyrolysis method with gas chromatography and a mass spectrometry was reported by Davison in 1954, and it became appreciated in recent years as a significant technique to analyze the three-dimensional network structure. This technique involves high-temperature treatment of the polymer membrane in order to isolate each element of the polymer using the solvent for gas chromatography, and to perform mass spectrometry of the gas chromatography peak at the same time.

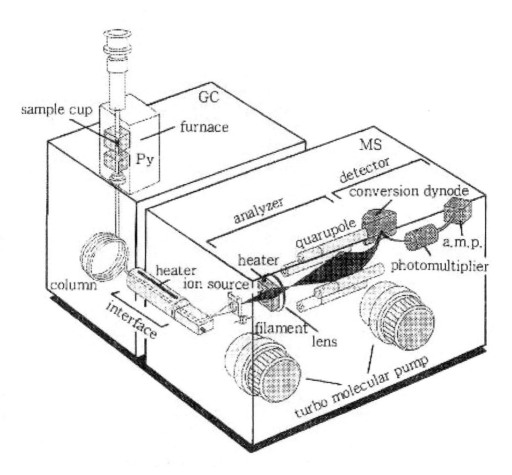

Fig. 2-1. Configuration of Py-GC/MS system

The advantage of this technique is the convenience of the device and pretreatment of the analysis sample. In particular, a polymer can be analyzed using a very small amount of sample (0.01-1mg).

The best feature of pyrolysis gas chromatography/mass spectrometry is pyrolysis. The thermal cracking unit can be heated quickly to a high temperature and is directly related to the reproducibility of the analysis performance. The heating in this method is done in a microfurnace (Figure 2-2).

The method involves putting the sample in the inactivated sample holder of the microfurnace, dropping it into the reactor core from the state set in the device, which is filled with helium as the carrier gas using a switch, and then pyrolyzing it. A feature of this technique is that the measurement result is steady because the small capacity keeps the temperature dispersion is comparatively low [21].

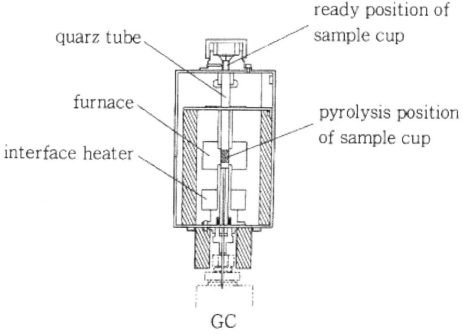

Fig. 2-2. Internal structure of pyrolyzer

2.1 Direct method (single-shot Py-GC/MS)

The easiest Py-GC/MS is a method in which only one heating performs the thermal decomposition, and it is called single shot. The range of temperatures that cause thermal decomposition is 50-1000°C (in the case of the PY-3030). After the sample is thermally decomposed, it was gasified and introduced into a gas chromatography. The gas separated by GC is measured by MS, as shown in Figure 2-3. The advantage of this technique is being able to analyze all components in the original sample by one measurement. This means that the ratio of components in the original solid can be measured.

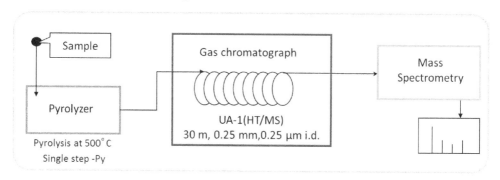

Fig. 2-3. Configuration of Py-GC/MS

The decomposition temperature of lacquer film is 500°C. The GC column is a type of dimethyl polysiloxane or phenyl methylpolysiloxane. Because lacquer film has a catechol configuration, its polarity is very high. Therefore, if a very polar column is used, detection will decrease. Moreover, setting the temperature of the oven high enough and using a heat-resistant column are also important because urushiol is undetectable at temperatures lower than 280°C.

2.2 Fragmentation mechanism of lacquer film

The main component of lacquer is urushiol, which is a catechol derivative, and it is polymerized to form a film by the laccase. When a lacquer film is analyzed by Py-GC/MS, mass fragment m/z=108 from alkyl phenol and mass fragment m/z=123 from alkyl catechol will appear as characteristic fragment ions. The fragment ion m/z=108 is the hydrogen attached to the 3-position carbon of the catechol ring that appears due to McLafferty transference. M/z=123 is a fragment ion which the first CH_2 from a catechol ring enters another catechol ring, and constitutes a seven-member ring, as shown in Figure 2-4.

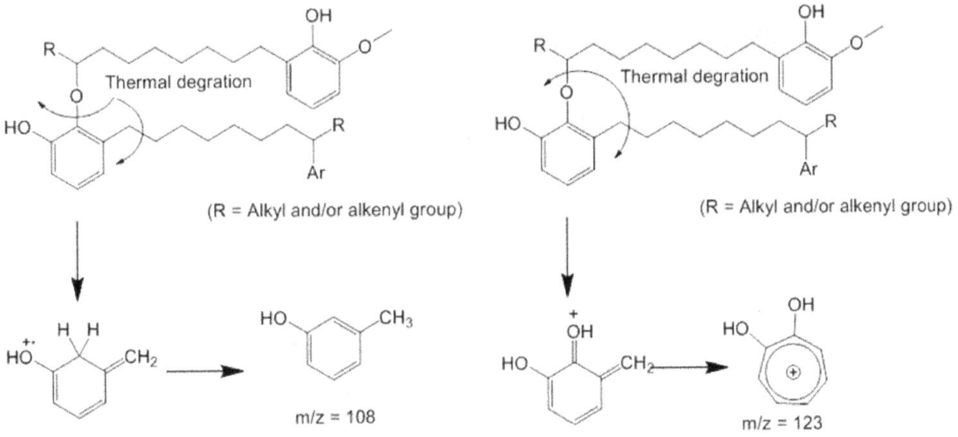

Fig. 2-4. Formation mechanism of fragment m/z = 108 and m/z = 123

The pyrolysis GC/MS Chinese lacquer film is shown in Figure 2-5. Many components are detected in the total ion chromatograph, and this is due to the complex structure of the lacquer film and various components contained in a lacquer sap. The ion chromatographs of (A) m/z=108 and (B) m/z=123 are shown in Figure 2-6. It is a graphic chart that is easily intelligible at a glance. The highest peak is the component that has carbon number seven in the side chain. Because the factor that becomes such is cut next to the 8th carbon in which most double bonds exist in urushiol. In the case of urushiol and thitsiol (Figure 2-7), carbon number seven is the most frequent component, but since the position of the double bond changes in laccol, the alkyl phenol of carbon number nine is detected as the greatest peak (Figure 2-8) [22-24].

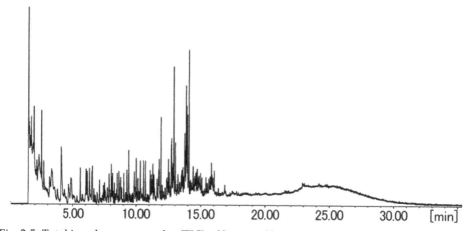

Fig. 2-5. Total ion chromatography (TIC) of lacquer film

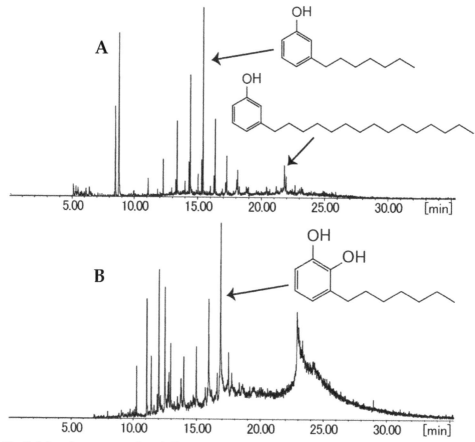

Fig. 2-6. Ion chromatography of Chinese lacquer film: (A) m/z=108 and (B) m/z=123

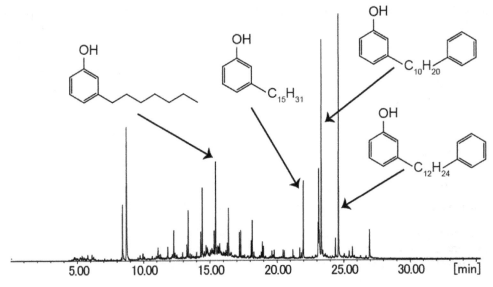

Fig. 2-7. Ion chromatography of Thai lacquer film

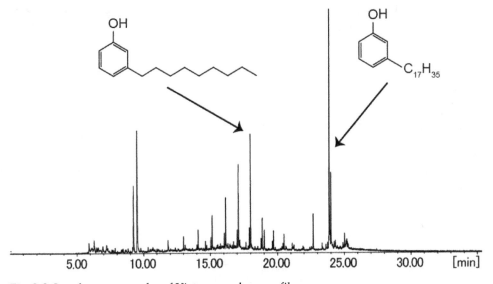

Fig. 2-8. Ion chromatography of Vietnamese lacquer film

2.3 Derivatization method

The principal constituent of Japanese lacquer, urushiol, contains catechol ring. Catechol has two hydroxyl groups of a phenol nature, and because its polarity is high, it tends to adhere to a column. It is said that this leads to a degradation of analytical sensitivity. Adsorption is one reason that the component ratios differ in analysis results. As a method of suppressing alterations of the component ratio by such adsorption, methylation of the hydroxyl group by tetramethylammonium hydroxide (TMAH) was devised. There are reports of the use of TMAH in analysis of a lignin, which generates a component containing the hydroxyl group during thermal decomposition and is an example of analysis of a polysaccharide [25-30]. TMAH can also be used effectively at the time of the thermal decomposition of Japanese lacquer. The lacquer film and TMAH were combined in the sample cup, and the sample cup was put in the pyrolyzer heated to 500°C.

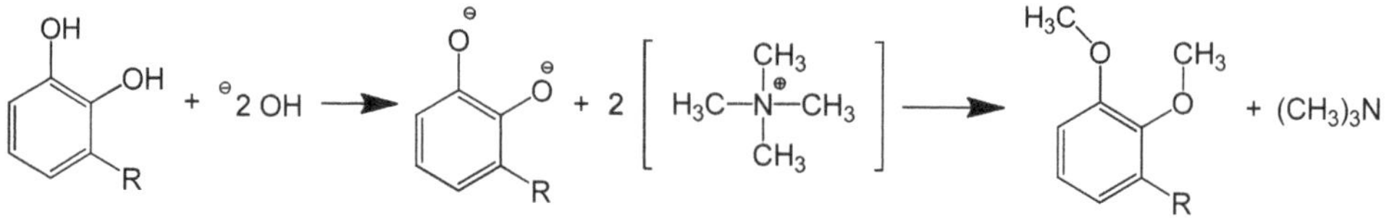

Scheme 2-1. Methylation reaction of hydroxyl group by TMAH

In a lacquer film, a hydroxyl group replaces a methyl group in a thermal decomposition (Scheme 2-1). The example of a Chinese lacquerware analyzed by this method is shown in Figure 2-9. In this examination, the amount of lacquer film used was 0.75 mg.

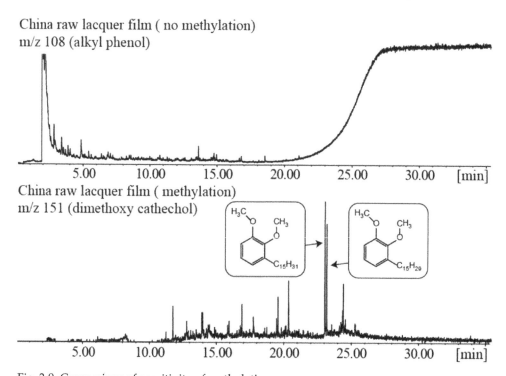

Fig. 2-9. Comparison of sensitivity of methylation

When not methylated, the alkylphenol of the carbon number seven should be detected. On the other hand, when it is methylated, 1,2-dimethoxy-3-pentadecylbenzene, from which urushiol was therefore protected by methylation, was detected. Methylation, even if the amount of lacquer film is small, can reveal urushiol. This method is useful for analysis of cultural properties that can provide only small samples.

2.4 Evolved gas analysis

Evolved gas analysis (EGA) is a method of analyzing the gas emitted during heat is applied to a sample [31-34]. The mechanical configuration characteristically uses a very short column. The column is about 1 m in length, and a fixed zone does not exist. Since many organic macromolecules are decomposed at less than 1000°C, a set of unstring temperatures in the range of 50–1000°C is used. In the case of a lacquer film, the temperature range is set to 50–650°C (Figure 2-10).

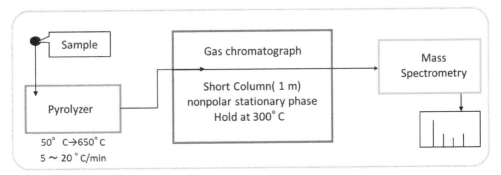

Fig. 2-10. Configuration of EGA system

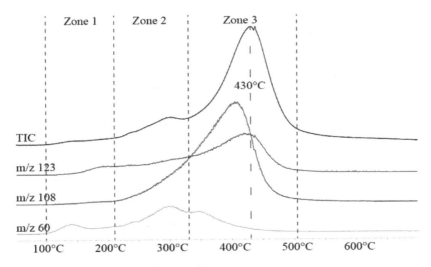

Fig. 2-11. EGA of Chinese lacquer film

EGA analysis decomposes a lacquer film in three steps (Figure 2-11). When the heat decomposition points were divided into three steps, two were the original peaks of m/z=60 and m/z=108, which is a thermal decomposition peak of urushiol, and one was m/z=123. The first peak of m/z=60 at a lower temperature was short carboxylic acid with various carbon chains. The following peak was considered to be due to the decomposition of sugars. The acetone powder (AP) taken from lacquer sap also was analyzed by EGA, and the peak position of acetone powder was mostly in agreement with the second EGA peak of Japanese lacquer film (Figure 2-12).

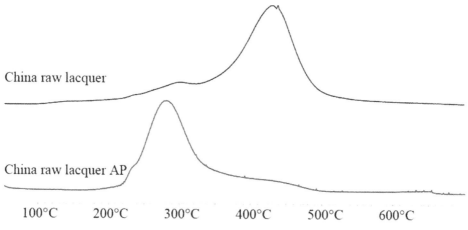

Fig. 2-12. EGA of Chinese lacquer film and AP

2.5 Double-shot pyrolysis

An analysis method that can be performed using the results of EGA is double-shot pyrolysis. Double-shot pyrolysis applies heat to the same sample gradually and analyzes only the gas generated in a certain range (Figure 2-13).

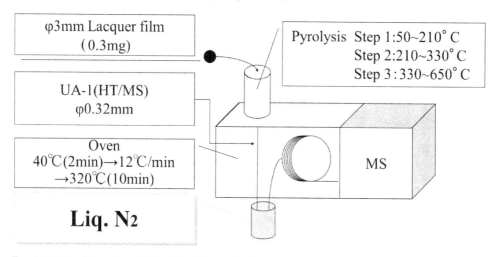

Fig. 2-13. Configuration of double-shot pyrolysis

A sample is placed in a pyrolyzer, and heating is started. The first part of a column is dipped in liquid nitrogen, and the gas chromatograph is not moved. If thermal decomposition occurs at a predetermined temperature, the pyrolyzed sample is removed from the heating block immediately. After raising the sample from the oven, removed the GC column from the liquid nitrogen and then started gas chromatography. Follow this procedure by the number of daylights of a thermal decomposition. In this way, the Py-GC/MS will store a record of each peak of EGA that is attained [35-37].

This technique is convenient to remove only a plasticizer previously added to a sample or to remove a solvent that remains in very small quantity. However, this method cannot be used when a sample reacts with the energy of heating. In the case of a lacquer sap, various additive admixtures (pine drying oil, tar, etc.) may be present. Thus, this technique is very effective when components with clearly different heat decomposition points are present.

3. Application of Py-GC/MS in lacquer analysis

The three kinds of lacquer films analyzed by Py-GC/MS in this section are the lacquer saps from *Rhus vernicifera*, *Rhus succedanea*, and *Melanorrhoea (Gluta) usitata*. The lacquers were coated on a 70 cm × 40 cm glass plates. They were allowed to polymerize in a humidity-controlled chamber at a relative humidity of 70-90% at 20°C for 12 h, and then removed from the chamber and stored in the open air for 3 years [38].

The pyrolysis-gas chromatography/mass spectrometry measurements were carried out using a vertical microfurnace-type pyrolyzer PY-2010D (Frontier Lab, Japan), an HP 6890 gas chromatograph (Hewlett-Packard), and an HPG 5972A mass spectrometer (Hewlett-Packard). A stainless steel capillary column (0.25 mm i.d. × 30 m) coated with 0.25 μm of Ultra Alloy PY-1 (100% methylsilicone) was used for the separation. The sample (1.0 mg) was placed in the platinum sample cup, and the cup was placed on top of the pyrolyzer at near ambient temperature. The sample cup was introduced into the furnace at 500°C, then the temperature program of the gas chromatograph oven was started. The gas chromatograph oven was programmed to provide a constant temperature increase of 20°C per min from 40° to 280°C, then held for 10 min at 280°C. The flow rate of the helium gas was 18 ml/min. All the pyrolysis products were identified by mass spectrometry. The mass spectrometry ionization energy was 70 eV (EI-mode).

3.1 Pyrolysis of urushiol polymer

The urushiol lacquer film was pyrolyzed at 500°C. The total ion chromatogram (TIC) and mass chromatogram (m/z=320) are shown in Figure 3-1. Although a complex TIC was obtained, the mass chromatogram and mass spectrum revealed that urushiol (MW=320) was the main component of the lacquer.

The pyrolysis products obtained in mass chromatograms and mass spectra of m/z=123 and m/z=108 of urushiol lacquer film are shown in Figure 3-2. 3-Heptylcatechol (C7) and 3-heptylphenol (C7) were detected in the mass chromatogram at m/z=123 and m/z=108, respectively, and had the highest relative intensity.

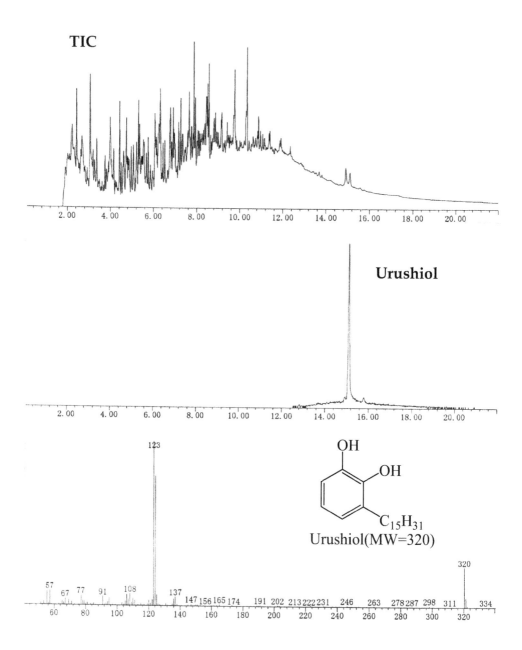

Fig. 3-1. TIC, mass chromatogram (m/z=320), and mass spectrum of urushiol lacquer film

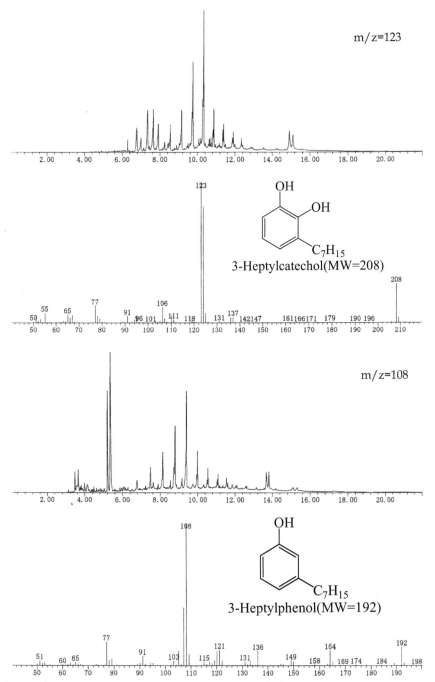

Fig. 3-2. Mass chromatograms (m/z=123 and 108) and mass spectra of urushiol lacquer film

It has been reported that the double bonds of olefins at the α- and β- positions are most susceptible to thermal cleavage [39-40]. Therefore, as shown in Scheme 3-1, the highest yield of 3-heptylcatechol (C7) can be attributed to the preferential cleavage at the α-position of the double bonds of the nucleus-14th chain C-O couplings of the urushiol polymers.

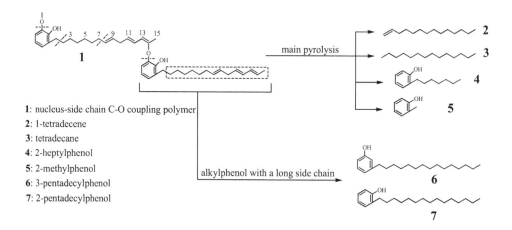

1: nucleus-side chain C-O coupling polymer
2: 1-tetradecene
3: tetradecane
4: 2-heptylphenol
5: 2-methylphenol
6: 3-pentadecylphenol
7: 2-pentadecylphenol

Scheme 3-1. Pyrolysis mechanism of urushiol polymer

3.2 Pyrolysis of laccol polymer

A laccol lacquer film was also pyrolyzed at 500°C at the same time as the urushiol lacquer film. The total ion chromatogram and mass chromatogram (m/z=320) are shown in Figure 3-3. Although a complex TIC was obtained, the mass chromatogram and mass spectrum confirmed that laccol (MW=348) was the main component of the lacquer film.

The pyrolysis products obtained in mass chromatograms and mass spectra of m/z=123 and m/z=108 of laccol lacquer film are shown in Figure 3-4. 3-Nonylcatechol (C9) and 3-nonylphenol (C9) were detected in the mass chromatogram at m/z=123 and m/z=108, respectively, and had the highest relative intensity. The detailed Py-GC/MS results of laccol in our laboratory have been previously reported [23].

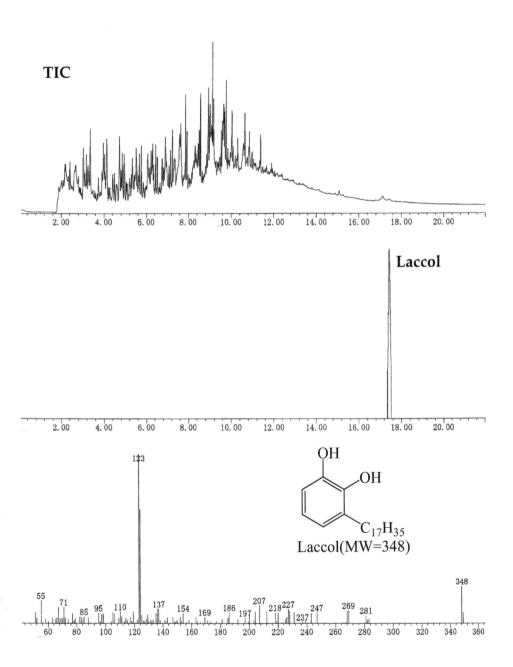

Fig. 3-3. TIC, mass chromatogram (m/z=348), and mass spectrum of laccol lacquer film

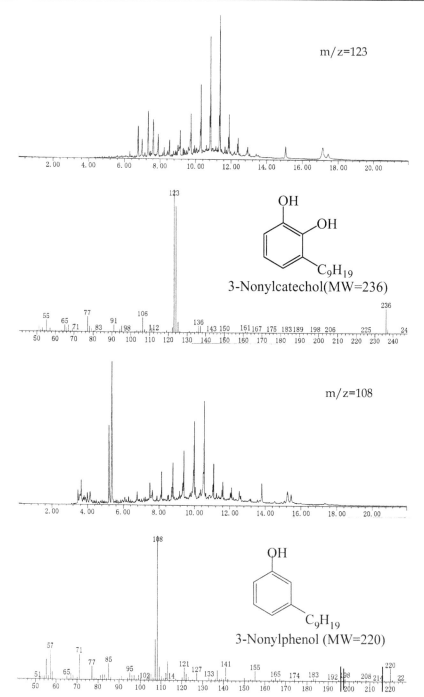

Fig. 3-4. Mass chromatograms (m/z=123 and 108) and mass spectra of laccol lacquer film

3.3 Pyrolysis of thitsiol polymer

Melanorrhoea (Gluta) usitata is a lacquer tree that grows in Myanmar, Thailand, and Cambodia. The main component of *Melanorrhoea (Gluta) usitata* is thitsiol, which contains 3- and 4-heptadecadienylcatechols as well as a series of α- and β- position.

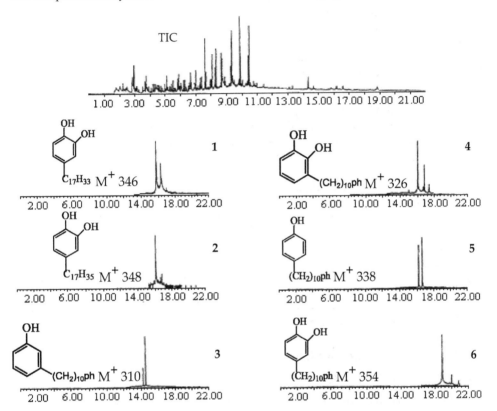

Fig. 3-5. TIC and mass chromatograms (m/z=346, 348, 310, 326, 338, 354) of thitsiol lacquer film

The thitsiol lacquer film was pyrolyzed at 500°C. Figure 3-5 shows the TIC and individual mass chromatograms at m/z=346, 348, 310, 326, 338, and 354. Peaks 1, 2, 3, 4, 5, and 6 of the mass chromatograms were identified as 4-hepatadecenylcatechol (MW=346), 4-hepatadecenylcatechol (MW=348), 3-(10-phenyldecyl)phenol (MW=310), 3-(10-phenyldecyl)catechol (MW=326), 4-(10-phenyldodecyl)phenol (MW=338), and 4-(12-phenyldodecyl)catechol (MW=354), respectively. Because the C-O coupling polymers are terminated with 4-hepatadecenylcatechol, these compounds can be formed from such terminal groups.

In the m/z=123 mass chromatograms of the thitsiol lacquer film, a pair of peaks of the 3- and 4-alkylcatechols were detected (figure not shown). The relative intensities of the 3- and 4-heptylcatechols (C7) were the highest in the pyrolysis products of the thitsiol lacquer. The highest yields of the 3- and 4-heptylcatechols were considered to be mainly due to cleavage at the α-position of the double bonds of the nucleus-8th and 12th chain C=O couplings for the thitsiol polymers. The alkylphenols detected are likely to be the products of pyrolysis of the nucleus-side chain coupling of the thitsiol polymers. Dimerization of the lacquer monomers is considered to proceed through the laccase-catalyzed nucleus-side chain coupling as well as the C-C coupling. The yields of 3-heptylphenol (C7) were the highest, as shown in Figure 3-5. The α- and β-positions of the double bonds of the olefins are susceptible to thermal cleavage so that these highest yields are thought to be produced primarily by cleavage at the α-position of the double bonds of thitsiol, such as the 3- and 4-(8,11-heptadecadienyl)catechols. The detailed Py-GC/MS results of thitsiol in our laboratory have been previously reported [41].

The 3-pentadecylcatechol (MW=320) (urushiol), 3-heptadecylcatechol (MW=348) (laccol), and 4-heptadecylcatechol (MW=348) (thitsiol) are the characteristic pyrolysis products of the three kinds of lacquer films, respectively, as summarized in Table 3-1. Furthermore, the data acquired for the alkylcatechols and alkylphenols in the pyrolysis products also can help to determine the lacquer species.

Species of lacquer film	Features of pyrolysis products		
	Monomer	Alkylcatechol	Alkylphenol
Rhus vernicifera	Urushiol	3-Heptylcatechol	3-Heptylphenol
Rhus succedanea	Laccol	3-Nonylcatechol	3-Nonylphenol
Melanorrhoea usitata	Thitsiol	3-Heptylcatechol, 4-Heptylcatechol	3-Heptylphenol

Table 3-1. Characteristic pyrolysis products of lacquer films

4. Recent finding on lacquer film using Py-GC/MS

4.1 Identification of lacquer species

As described in Section 3, the main pyrolysis products of urushiol are 3-heptylcatechol and 3-heptylphenol, of laccol are 3-nonylcatechol and 3-nonylphenol, and of thitsiol are 3-heptylcatechol, 3-heptylphenol, and 4-heptylcatechol, respectively. In order to confirm these results, we synthesized urushiol [42], laccol [23], and thitsiol [41] lacquer films and analyzed them by Py-GC/MS measurement.

4.1.1 Synthesis of urushiol lacquer film and analysis by Py-GC/MS

Compound **1**, the major component of urushiol, shown in Figure 4-1, was synthesized in good yield via the Witting reaction from a ylide derived from (3E,5Z)-3,5-heptadienyltriphenylphosphonium iodide with 3-(8-oxo-1-octyl) catechol diacetate, using a stepwise procedure based on repeated protection and deprotection of the hydroxyl group of catechol, using the technique we reported previously [43]. The trienyl urushiol compounds **2** and **3** were synthesized in a similar method using ylides derived from (3E,5E)-3,5-heptadienyltriphenylphosphonium iodide and 3E-3,5-heptadienyltriphenylphosphonium iodide, respectively.

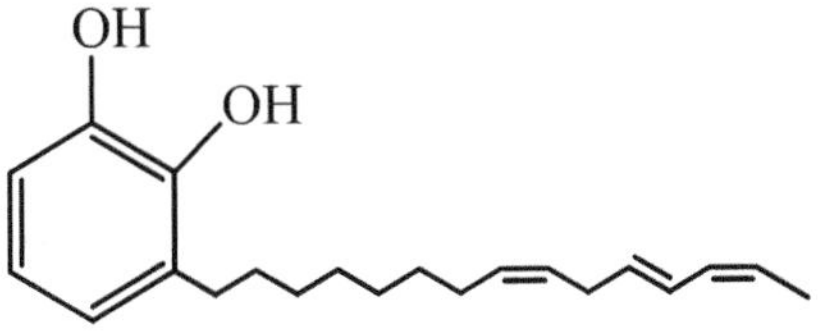

1

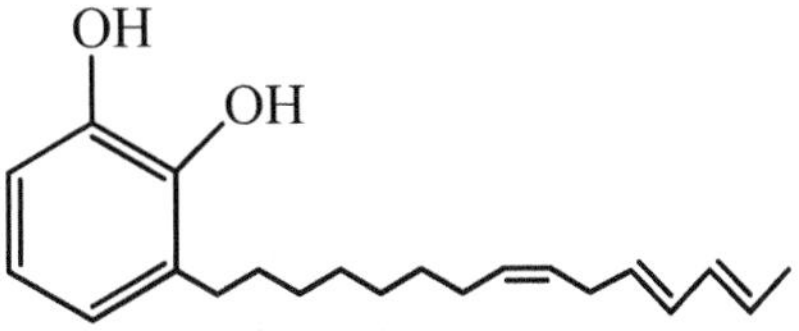

2

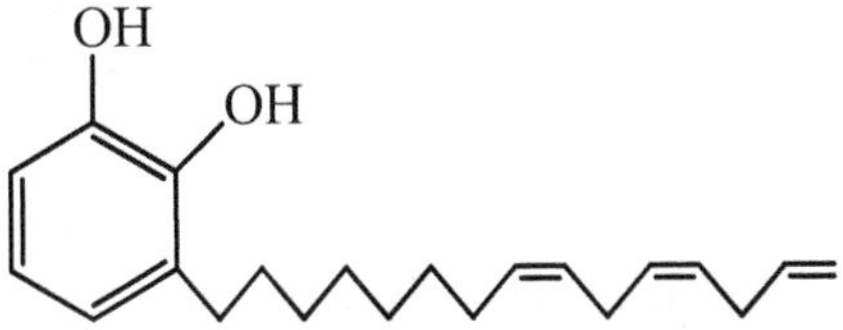

3

Fig. 4-1. Stereo-structure of trienyl urushiols: (1) 3-[(8Z,11E,13Z)-8,11,13-pentadecatrienyl]catechol, (2) 3-[(8Z,11E,13E)-8,11,13-pentadecatrienyl]catechol, (3) 3-[(8Z,11E)-8,11,14-pentadecatrienyl]catechol

Synthesized lacquer films were prepared as follows: 50 mg of compound 1 was added to 10 mg water-isopropyl alcohol (1:1,v/v) containing acetone powder (10 mg), which had been separated as an acetone-insoluble material from *Rhus vernicifera* lacquer sap. The resulting mixture was stirred for about 15 min. The reaction mixture had a viscosity suitable for coating and was coated onto a glass plate and dried in a humidity-controlled chamber with a relative humidity (RH) of 80% at 25–30°C for 10 h. The film was then removed from the chamber and stored in air for 6 months. Films of a single component were prepared. Compounds 2 and 3 were treated in the same manner to obtain the corresponding lacquer films. The films from the synthesized trienyl urushiol components 1, 2, and 3 are called synthesized lacquer films A, B, and C, respectively.

The preparation of natural lacquer was as follows. The sap exuded from a *Rhus vernicifera* lacquer tree in Japan, and composed of the following lipid components: 5% saturated urushiol, 18% monoenyl urushiol, 12% dienyl urushiol, and 65% trienyl urushiol. This lacquer sap was coated onto a glass plate and dried in a humidity-controlled chamber under an RH of 80% at 25–30°C for 10 h. The film was then removed from the chamber and stored in air for 6 months like the synthesized lacquer films.

The three synthesized lacquer films and the natural lacquer film were pyrolyzed at 500°C, and then the resulting pyrolysis products were characterized by GC-MS analysis. The specific ions at m/z 123 and 108 are fragment ions of alkylcatechols and alkylphenols that were produced during the electron ionization process in the mass spectrometer, and the mechanism of the thermal decomposition is shown in Figure 4-2.

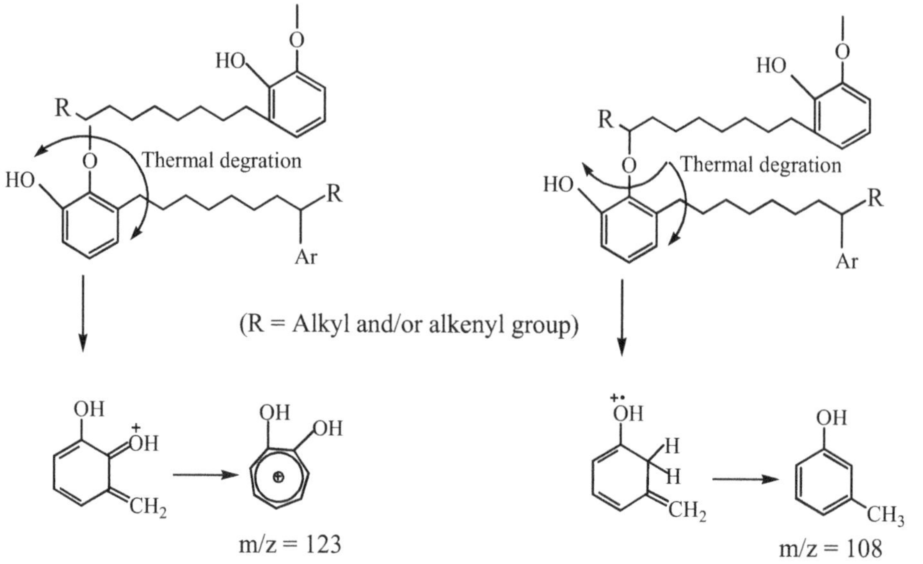

Fig. 4-2. Postulated location of urushiol lacquer film pyrolysis and mechanism of formation of m/z=123 and m/z=108 ion species in Py-GC/MS

The resulting total ion chromatograms (TIC) of the natural and synthesized lacquer films are shown in Figure 4-3. The major components in the TIC peaks were defined as alkanes, alkenes, and alkylbenzenes respectively.

The mass chromatogram of selective scanning of m/z=123 is shown in Figure 4-4. Peaks 1, 2, and 3 were identified as 3-pentylcatechol (M^+=180), 3-hexylcatechol (M^+=194), and 3-heptylcatechol (M^+=208), respectively. Peaks 4 and 5 were observed in only the natural lacquer film, and were identified as the urushiol monomer components 3-pentadecenylcatechol (M^+=318) and 3-pentadecylcatechol (M^+=320), based on analysis of their mass spectra, shown in Figure 4-5.

Figure 4-5 shows the mass spectra of compounds giving rise to peaks 1–5 in the selective plotting of m/z=123 ion species in the spectra from the pyrolysis of synthesized and natural lacquer films shown in Figure 4-4. Thus, it is evident from Figure 4-4 that saturated urushiol (peak 5) and monoenyl urushiol (peak 4) were detected by pyrolysis of the natural lacquer film, but not of the synthesized lacquer films. These urushiols likely produced thermally decomposed fragments from the terminal alkylcatechol and alkenylcatechol side chains of the natural lacquer film. Therefore, it was confirmed that the natural lacquer film has urushiol components with both saturated and monoenyl side chains, unlike the synthesized lacquer films.

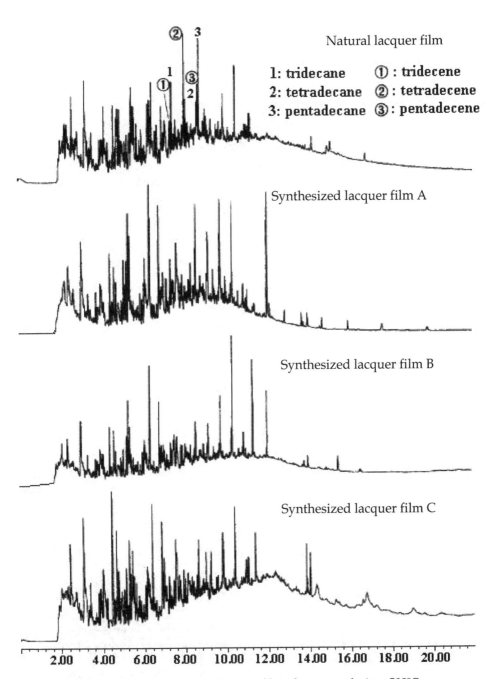

Fig. 4-3. TIC of natural and synthesized lacquer films due to pyrolysis at 500°C

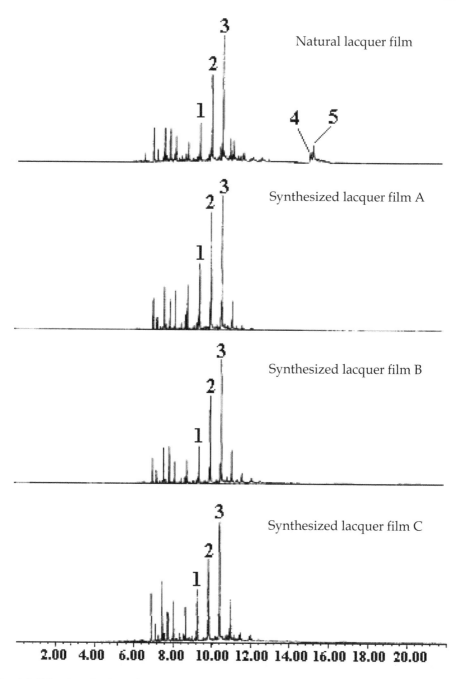

Fig. 4-4. Selective plotting of ion species (m/z=123) in spectra obtained from TIC of one natural and three synthesized lacquer films

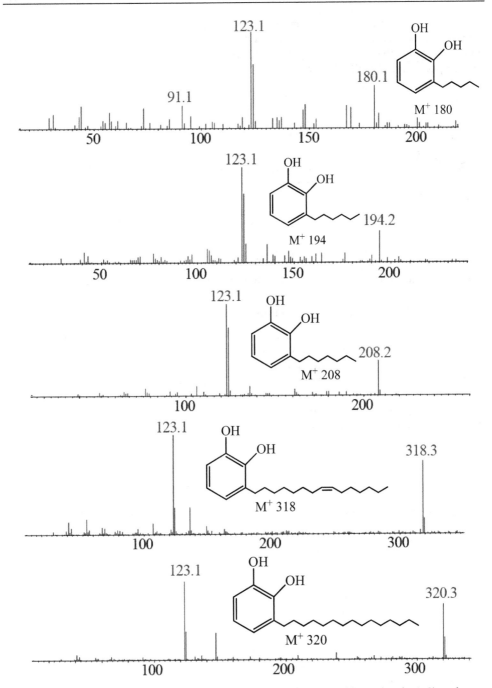

Fig. 4-5. Mass spectra of alkylcatechols from synthesized lacquer film A (peaks 1–3) and natural lacquer film (peaks 4–5) in mass chromatogram (m/z=123)

The highest peak intensity in each mass chromatogram was peak 3, because most urushiol components have a double bond at position 8 of their side chain, and the bond between C7 and C8 is easily thermally decomposed after polymerization as shown in Figure 4-6. The highest peak intensity of *Rhus vernicifera* was 3-heptylcatechol (M^+=208).

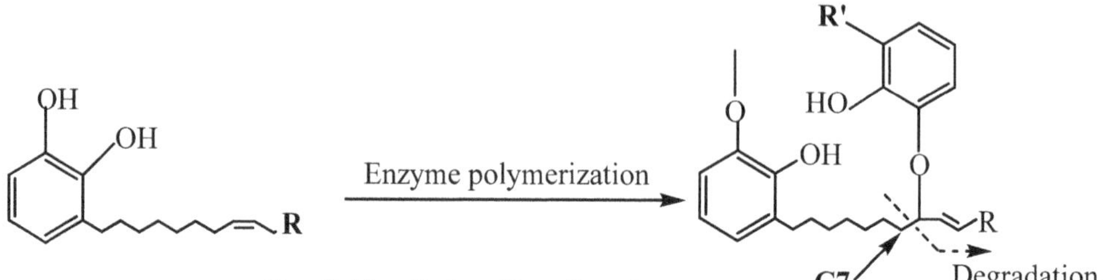

$R = C_5H_n$, alkyl and/or alkenyl group

$R' = C_{15}H_m$, alkyl and/or alkenyl group

Fig. 4-6. Postulated location of lacquer film in thermal decomposition due to pyrolysis

4.1.2 Synthesis of laccol lacquer film and analysis by Py-GC/MS

The trienyl and dienyl laccols, compounds 4, 5, and 6, shown in Figure 4-7 were synthesized by a method similar to the synthesis of urushiol described in Section 4.1.1. Compound 4, the major component of laccol, was synthesized via the Witting reaction from a ylide derived from (3E,5E)-3,5-heptadienyltriphenylphosphonium iodide with 3-(8-oxo-1-decyl)catechol diacetate. Compounds 5 and 6 were synthesized from 3Z-heptadienyltriphenylphosphonium iodide and 3E-heptadienyltriphenylphosphonium iodide, respectively.

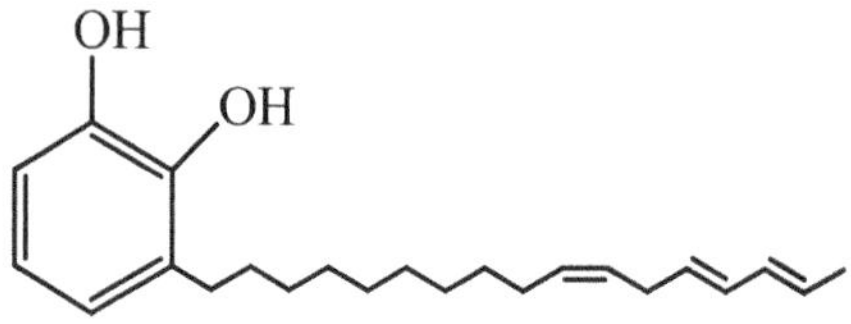

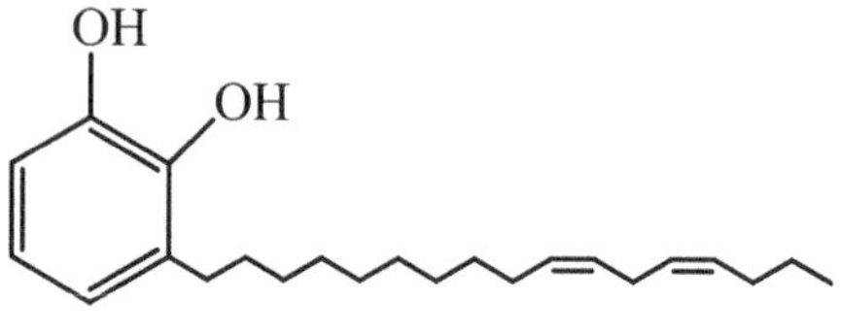

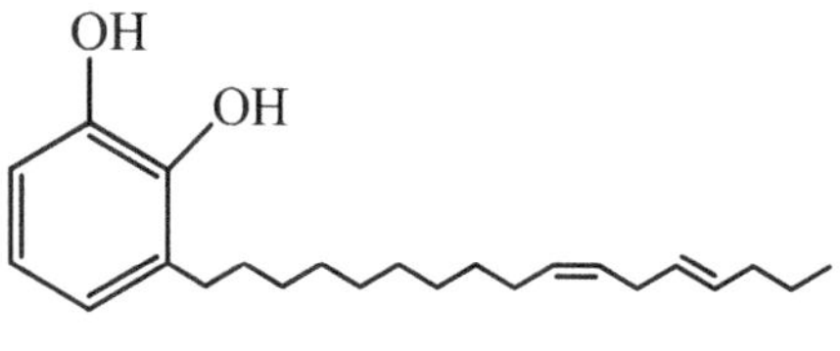

Fig. 4-7. Stereo-structure of trienyl and dienyl laccols: (4) 3-[(10Z,13E,15E)-10,13,15-heptadecatrienyl]catechol, (5) 3-[(10Z,13Z)-10,13-heptadecadienyl]catechol, (6) 3-[(10Z,13E)-10,13-heptadecadienyl]catechol

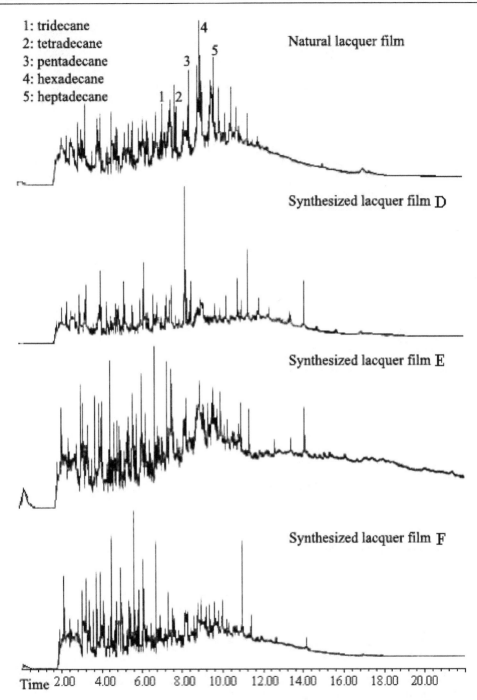

Fig. 4-8. TIC of natural and synthesized laccol lacquer films due to pyrolysis at 500°C

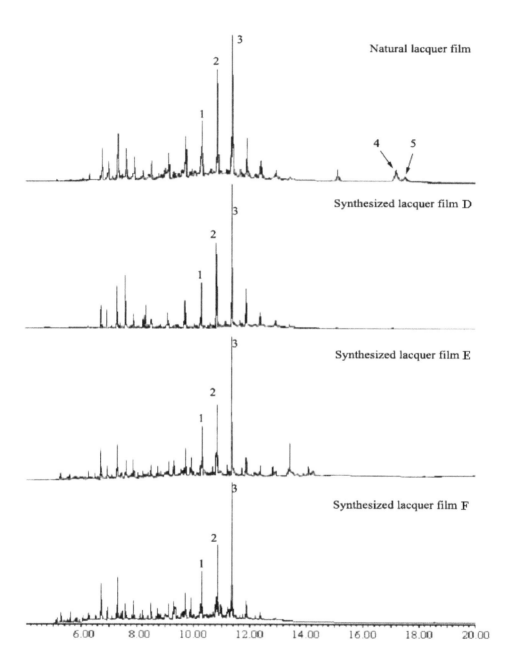

Fig. 4-9. Selective plotting of ion species (m/z=123) in spectra obtained from TIC of one natural and three synthesized lacquer films

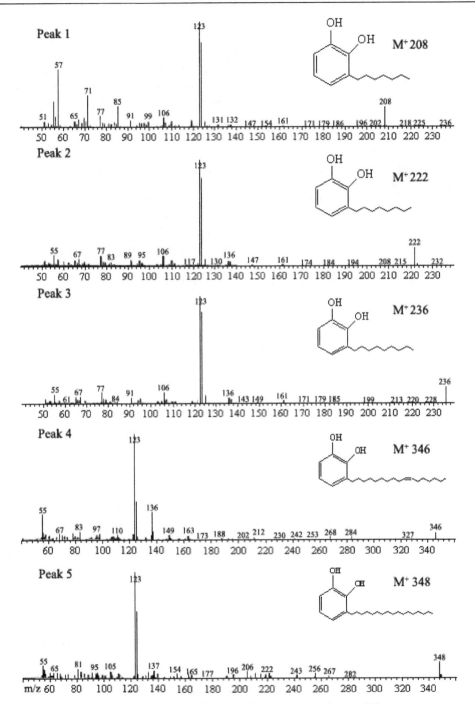

Fig. 4-10. Mass spectra of peaks 1–5 in the mass chromatogram (m/z=123)

Synthesized and natural laccol lacquer films also were prepared by methods similar to those of urushiol lacquer film described in Section 4.1.1. However, because the laccase activity of *Rhus succedanea* is lower than that of *Rhus vernicifera*, the laccol lacquer films were dried in a humidity-controlled chamber with RH of 90% at 30°C for 10 h. Then they were removed from the chamber and stored in air for 6 months. The films from the synthesized laccol components 4, 5, and 6 are called synthesized lacquer films D, E, and F, respectively.

The three synthesized laccol lacquer films and one natural *Rhus succedanea* lacquer film were pyrolyzed at 500°C, and the resulting pyrolysis products were characterized by GC-MS analysis. The specific ions at m/z 123 and 108 are fragment ions of alkylcatechols and alkylphenols that were produced during the electron ionization process in the mass spectrometer, like those of the urushiol lacquer film. The TIC of natural and synthesized laccol lacquer films are shown in Figure 4-8. The major components in the TIC peaks were defined as alkanes, alkenes, and alkylbenzenes, respectively.

The mass chromatogram of the selective scanning of m/z=123 shown in Figure 4-9. Peaks 1, 2, and 3 were 3-heptylcatechol (M^+=208), 3-octylcatechol (M^+=222), and 3-nonylcatechol (M^+=236) respectively. Peaks 4 and 5 were observed in only the natural lacquer film, and were identified as laccol monomer components, 3-heptadecenylcatechol (M^+=346) and 3-heptadecylcatechol (M^+=348), based on their mass spectra (Figure 4-10).

These laccols were likely produced as thermally decomposed fragments from the terminal alkyl and alkenylcatechol of the natural *Rhus succedanea* lacquer film. Therefore, the natural lacquer sap has laccol components with both saturated and monoenyl side chains, but the synthesized laccols did not. This was the same result as that of the natural lacquer film from *Rhus vernicifera*. The highest peak intensity was peak 3 in each mass chromatogram because most laccol components have a double bond at position 10 of their side chain, and the bond between C9 and C10 is easily thermally decomposed after polymerization. This result was different from that of the lacquer film from *Rhus vernicifera*, which has a double bond at position 8 of its side chain. The highest peak intensity of *Rhus vernicifera* was 3-heptylcatechol (M^+=208), of *Rhus succedanea* was 3-nonylcatechol (M^+=236).

4.1.3 Synthesis of thitsiol lacquer film and analysis by Py-GC/MS

The main component of thitsiol, 3-(10-phenyldecyl)catechol, was synthesized by the reaction of dimethyl ether with catechol and 1-phenyl-10-iododecane with n-BuLi, followed by

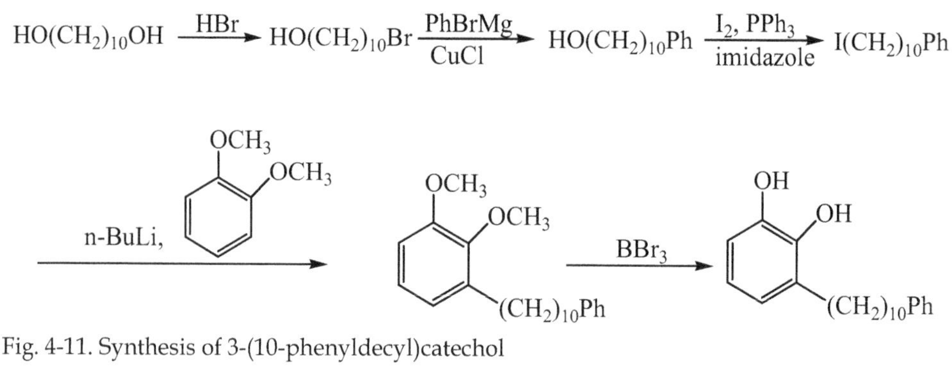

$$HO(CH_2)_{10}OH \xrightarrow{HBr} HO(CH_2)_{10}Br \xrightarrow[CuCl]{PhBrMg} HO(CH_2)_{10}Ph \xrightarrow[imidazole]{I_2,\ PPh_3} I(CH_2)_{10}Ph$$

Fig. 4-11. Synthesis of 3-(10-phenyldecyl)catechol

deprotection of the hydroxyl groups of catechol, as shown in Figure 4-11. The detail experimental process can be found in the literature (3 and 4).

The preparation of synthesized thitsiol lacquer films is as follow: 50 mg 3-(10-phenyldecyl)catechol was added to 50 mg of a water-isopropyl alcohol mixture (1:1,v/v) containing 20 mg acetone powder obtained from a natural *Melanorrhoea (Gluta) usitata* lacquer sap. The resulting mixture was stirred for about 15 min, and then applied to a glass sheet and dried in a humidity-controlled chamber with a RH of 90% at 30°C for 10 h. The film was then removed from the chamber and stored in air for 6 months. The natural lacquer film also was prepared by the same method.

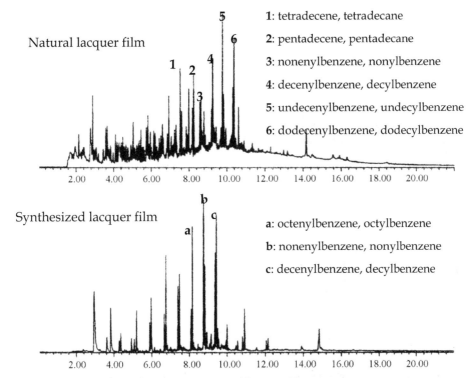

Fig. 4-12. TIC of natural and synthesized thitsiol lacquer films

The synthesized thitsiol lacquer films and a natural *Melanorrhoea (Gluta) usitata* lacquer film were pyrolyzed at 500°C, and then the resulting pyrolysis products were characterized by GC-MS analysis. Figure 4-12 shows the TIC of the natural and synthesized lacquer films. The main pyrolysis products of the natural lacquer film were identified as alkanes, alkenes, alkylbenzenes and alkenylbenzenes by the mass spectra. On the other hand, alkanes and alkenes were not detected in the TIC of the synthesized lacquer film because the alkanes and alkenes were derived from 3- or 4-alkenylcatechols contained in the natural lacquer sap. Each peak in TIC is a pair of peaks of saturated and unsaturated components having one double bond in the termination. It was considered that the alkanes and alkenes were formed by decomposition of the terminal of the lipid components in the polymer.

Figure 4-13 is the mass chromatograms (m/z=123) of natural and synthesized thitsiol lacquer films. Alkylcatechol was detected in the mass chromatogram of the natural lacquer film, and detected as a pair peak of 3-alkylcatechol and 4-alkylcatechol. Due to the analysis of each mass spectrum, the detailed assignment was as follows: ①: 3-methylcatechol, 1: 4-methylcatechol; ②: 3-ethylcatechol, 2: 4-ethylcatechol; ③: 3-propylcatechol, 3: 4-propylcatechol; ④: 3-butylcatechol, 4: 4-butylcatechol; ⑤: 3-pentylcatechol, 5: 4-pentylcatechol; ⑥: 3-hexylcatechol, 6: 4-hexylcatechol; ⑦: 3-heptylcatechol, 7: 4-heptylcatechol; ⑧: 3-octylcatechol, 8: 4-octylcatechol; 9: 3-pentadecylcatechol; 10: 3-(10-phenyldecyl)catechol.

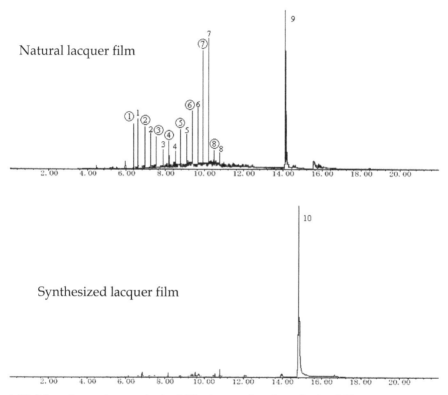

Fig. 4-13. Mass chromatogram (m/z=123) of natural and synthesized films

Figure 4-14 shows the mass spectra of pair of peaks ⑥-6 and ⑦-7. These components were derived from 3-alkylcatechol or 4-alkylcatechol contained as monomeric components in *Melanorrhoea (Gluta) usitata*. On the other hand, alkylcatechol was not detected in the mass chromatogram of the synthesized lacquer film because the synthesized lacquer film included no alkylcatechols as monomeric components. The side chain of the polymerized ω-alkylcatechol was difficult to decompose by heating because ω-alkylcatechol is not polymerized at the side chain by auto-oxidation like a fatty acid. Therefore, alkylcatechol was not detected in the mass chromatogram. However, 3-(10-phenyldecyl)catechol was detected as a monomer. The highest pair-peaks intensity of *Melanorrhoea (Gluta) usitata* was 3-heptylcatechol and 4- heptylcatechol (M+=208, peaks ⑦ and 7 in Figure 4-13).

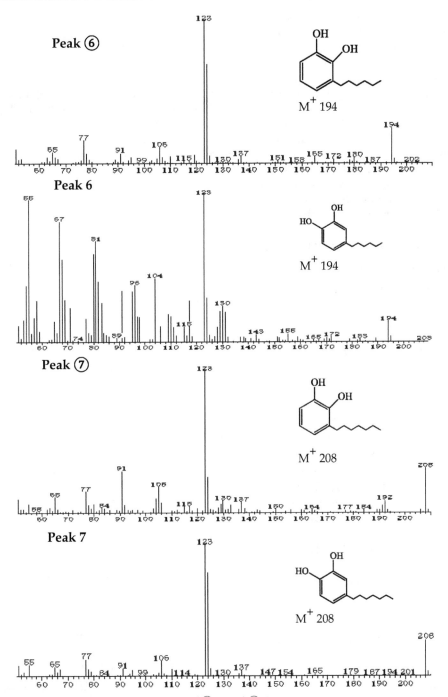

Fig. 4-14. Mass spectra of pair of peaks of ⑥-6 and ⑦-7.

The synthesized lacquer films and corresponding natural lacquer films were characterized using Py-GC/MS. The saturated and monoenyl components were detected in the natural *Rhus vernicifera* and *Rhus succedanea* lacquer films, respectively, but not in the synthesized lacquer films. However, alkylphenols, alkenylphenols, alkanes, and alkenes having longer carbon chains than the side chains of the synthesized lacquers were detected by the Py-GC/MS analysis. Comparing the peak intensity of the mass chromatograms with that of *Rhus succedanea*, *Rhus vernicifera*, and *Melanorrhoea (Gluta) usitata* lacquer films showed that the highest peak of *Rhus vernicifera* represented the C7 components (heptylcatechol and heptylphenol), and the highest peak of *Rhus succedanea* represented the C9 components (nonylcatechol and nonylphenol) due to a double bond in position 10 of its side chain of laccol. On the other hand, a pair of peaks of 3-alkylcatechol and 4-alkylcatechol was detected in the *Melanorrhoea (Gluta) usitata* lacquer film. The highest pair peaks were 3-heptylcatechol and 4-heptylcatechol due to the side chain attached in position 3 and/or position 4 of catechol ring of thitsiol.

4.2 Identification of ancient lacquerware

Py-GC/MS is a powerful and versatile technique to identify Oriental lacquers as described above. In this section, we describe analysis of several ancient lacquerwares by Py-GC/MS, and compared with the results of natural lacquer films to determine the kind of lacquer.

4.2.1 Ryukyu lacquerware

We have analyzed six kinds of Ryukyu lacquerware by Py-GC/MS to determine the lacquer source [44]. Six pieces of lacquer belonging to Urasoe Art Museum are summarized in Table 4-1, and the photos of each lacquerware object are shown in Photo 4-1. Each sample was removed from the ancient lacquerware objects during restoration. All pieces of Ryukyu lacquerware in Table 4-1 were analyzed by pyrolysis-gas chromatography/mass spectrometry at 500°C.

No.	Collection method	Art object (Museum I.D. number)
1	Naturally flaked off	Black lacquer wood box with *Raden* (43)
2	Naturally flaked off	Black lacquer table box with *Raden* (66)
3	Naturally flaked off	Black lacquer box with *Hakue* (220)
4	Naturally flaked off	Black lacquer box with *Hakue* (416)
5	Collection during restored	Black lacquer box with *Raden* (26)
6	Collection during restored	Red lacquer fabric bowl with *Hakue* (100)

Table 4-1. Samples used for pyrolysis

No. 1   No. 2   No. 3

No. 4   No. 5   No. 6

Photo 4-1. Six lacquer objects belonging to Urasoe Art Museum

Figure 4-15 shows the TIC and mass chromatogram (m/z=320, m/z=123, m/z=108) of Sample 2. The peak at the retention time of 14.326 min (m/z=320) is urushiol, at 10.038 min (m/z=123) is 3-heptylcatechol, and at 9.139 min (m/z=108) is 3-heptylphenol respectively. The mass spectra are shown in Figure 4-16. Samples 1, 4, and 5 showed TIC, mass chromatograms, and mass spectra similar to that of Sample 2, suggested that these four lacquerwares were made from *Rhus vernicifera*.

The TIC and mass chromatogram of Sample 3 are shown in Figure 4-15, and the mass spectrum is shown in Figure 4-16. The results showed that the peak at the retention time of 16.072 min (m/z=348) is laccol, at 11.105 min (m/z=123) is 3-nonylcatechol, and at 10.282 min (m/z=108) is 3-nonylphenol. Sample 6 produced the same pyrolysis results as Sample 3, suggesting that these two lacquerwares were made from *Rhus succedanea*. The pyrolysis products of the six samples are summarized in Table 4-2.

The Py-GC/MS results of Ryukyu ancient lacquerware confirmed that four lacquerware objects belonging to the Urasoe Art Museum were made from lacquer sap of *Rhus vernicifera* and the other two were made from lacquer sap of *Rhus succedanea* lacquer sap. Because Py-GC/MS analysis revealed the answer to the question of what kind of lacquer sap was used to produce Ryukyu lacquerware, the procurement source and production system of the materials of Ryukyu lacquerware could be clarified, and will be very useful in the conservation and restoration of other valuable ancient lacquer ware.

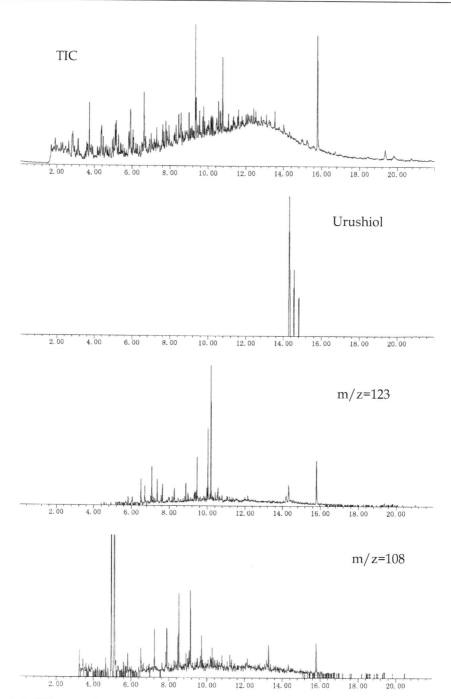

Fig. 4-15. TIC, mass chromatograms (m/z=320, 123, and 108) of sample 2.

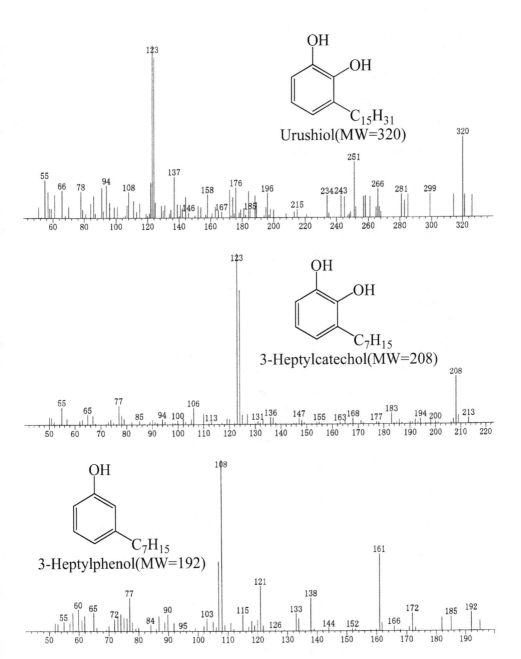

Fig. 4-16. Mass spectra of sample 2

No.	Lacquer species	Pyrolysis products		
		Monomer	Alkylcatechol	Alkylphenol
1	*Rhus vernicifera*	Urushiol	3-heptylcatechol	3-heptylphenol
2	*Rhus vernicifera*	Urushiol	3-heptylcatechol	3-heptylphenol
3	*Rhus succedanea*	Laccol	3-nonylcatechol	3-nonylphenol
4	*Rhus vernicifera*	Urushiol	3-heptylcatechol	3-heptylphenol
5	*Rhus vernicifera*	Urushiol	3-heptylcatechol	3-heptylphenol
6	*Rhus succedanea*	Laccol	3-nonylcatechol	3-nonylphenol

Table 4-2. Pyrolysis products of Ryukyu lacquer films and their species

4.2.2 Other ancient lacquerware

An ancient lacquer film was obtained from the surface of a wooden dish extracted from an excavation site that dates back to the 17th–18th century A.D. at *Kinenkanmae Iseki* on the campus of Meiji University in Tokyo, Japan, called Sample 1. A *Nanban* lacquer film from the 17th century A.D., called Sample 2, and an old lacquer film imported from an Asian country during the 17th–18th century A.D., called Sample 3, were obtained from the surface of wooden crafts objects [38]. Pieces of lacquer taken from a four-eared jar that is a Japanese National Important Cultural Property found in 16th–17th century Kyoto ruins, called Sample 4, was a piece of lacquer obtained from the side of the vessel [22].

The *Baroque* and *Rococo* lacquer films were obtained from the wood surfaces of the Rococo church St. Alto in Altomunster, Munich, Germany, identified as Samples 5 [38]. It is sometimes extremely important to determine whether objects that are claimed to be lacquerware are actually created from lacquer sap or other resins. Whether an object is lacquerware can be precisely determined by the presence of urushiol, laccol, or thitsiol with alkylcatechols and alkylphenols using Py-GC/MS.

The TIC and mass chromatograms of m/z=320 of Sample 1 are shown in Figure 4-17. Urushiol, 3-pentadecylcatechol (MW=320), was identified as the monomer of the lacquer film based on the mass spectrum and retention time. This result was compared to those of the three types of Oriental lacquers. 3-Pentadecylcatechol of MW=320 is the saturated urushiol component, which is the monomer of the *Rhus vernicifera* lacquer. The monomers of the *Rhus succedanea* lacquer are laccol components such as 3-pentadecylcatechol of MW=348. This was not detected in the Sample 1 except for the 3-pentadecylcatechol of MW=320. The monomer of *Melanorrhoea usitata* lacquer is thitsiol, which has saturated and monoenyl side chains, such as 4-hepatadecylcatechol (MW=348) and 3- and 4-(ω-phenylalkyl) phenols and catechols, and these components, except for the 3-pentadecylcatechol of MW=320, were not detected in the Sample 1. It was concluded that the Sample 1 lacquer was made from *Rhus vernicifera* lacquer sap.

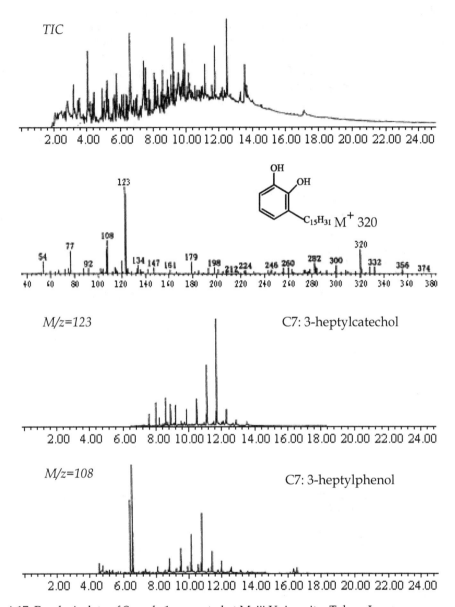

Fig. 4-17. Pyrolysis data of Sample 1 excavated at Meiji University, Tokyo, Japan

The TIC and mass chromatograms of m/z=60, m/z=123, and m/z=108 from the pyrolysis products of the Sample 2 are shown in Figure 18. The mass chromatograms of m/z=60 indicated that the Sample 2 lacquerware included a drying oil [45], which was added to retard the rate of hardening and affected the physical properties of the film. The mass chromatograms of m/z=108 and m/z=123 of the pyrolysis products of the lacquerware are also shown in Figure 4-18. The greatest number of peak were from 3-heptylphenol (C7) and 3-heptylcatechol (C7), as revealed by the mass spectra. It was concluded that the Sample 2 lacquer was made from *Rhus vernicifera* lacquer sap. Urushiol was not detected because the surface of the Sample 2 lacquerware was oxidized by oxygen and light. A type of wax was detected in the mass spectrum of the TIC from the pyrolysis products. It is considered that the wax was used to polish the surface of the lacquerware.

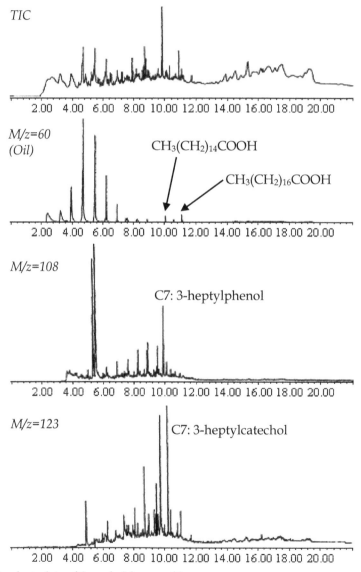

Fig. 4-18. Pyrolysis data of Sample 2 lacquer film

The TIC and mass chromatograms of m/z=123 and m/z=108 for the pyrolysis products of the Sample 3 are shown in Figure 4-19. Alkylcatechols and alkylphenols were detected in the mass chromatograms. The greatest number of peaks was due to the 3- and 4-nonylcatechols (C9) and 3-heptylphenol (C7), as determined from the mass spectra. It was concluded that the Sample 3 lacquerware was produced using *Melanorrhoea (Gluta) usitata* lacquer sap. Thitsiol was not detected in the pyrolysis data because the surface of the lacquerware was oxidized by oxygen and exposure to light. From the TIC and mass spectra of the pyrolysis products, a type of wax was detected. The wax was determined to be a type of beeswax, as revealed by the mass spectra. The beeswax was used to protect and polish the surface of the lacquerware.

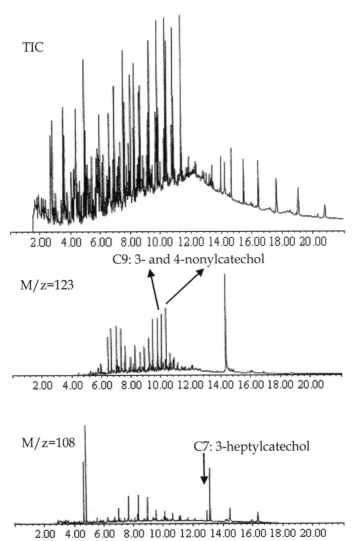

Fig. 4-19. Pyrolysis data of Sample 3

The TIC and mass chromatograms of m/z=123 and m/z=108 for the pyrolysis products of Sample 4 are shown in Figure 4-20. The peak at the retention time of 17.3 min (*m/z* = 123, peak 1) is 3-(10-phenyldecyl)catechol, and at 20.4 min (*m/z* = 123, peak 2) is 3-(10-phenyldodecyl)catechol, according to the results of analysis of the mass chromatogram. It was concluded that the Sample 4 lacquer was made from *Melanorrhoea (Gluta) usitata* lacquer sap.

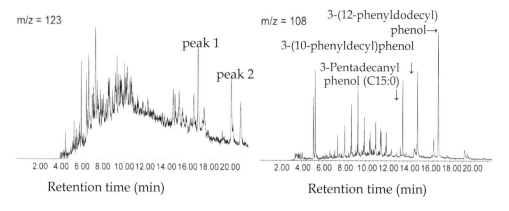

Fig. 4-20. Mass chromatograms of Sample 4

Figure 4-21 shows the Py-GC/MS products of the Sample 5 (Baroque and Rococo) lacquer films. Lacquer components were not detected in the TIC on the mass chromatograms of the alkylcatechols of m/z=123 and alkylphenols of m/z=108 of the pyrolysis products. Monoterpene components and sesquiterpene components were detected in the pyrolysis products of both lacquerwares. It was concluded that the lacquerwares were made from natural resins, but the pyrolysis products of the Baroque and Rococo lacquer and natural resins using this method were not clear.

It was concluded that 3-pentadecylcatechol (MW=320) (urushiol), 3-heptadecylcatechol (MW=348) (laccol), and 4-heptadecylcatechol (MW=348) (thitsiol) are the main products of the pyrolysis of *Rhus vernicifera*, *Rhus succedanea*, and *Melanorrhoea usitata*, respectively. Compared with the results of the natural lacquer film, the ancient lacquer film (Sample 1) and *Nanban* lacquer film (Sample 2) were assigned to *Rhus vernicifera*, both the old lacquerware objects imported from an Asian country (Sample 3) and the lacquer taken from a four-eared jar that is a Japanese National Important Cultural Property obtained from 16th–17th century Kyoto ruins (Sample 4) were assigned to *Melanorrhoea (Gluta) usitata*. However, although they were also called "lacquer," the Baroque and Rococo (Sample 5) lacquer film were identified as being made from a natural resin. The pyrolysis products clearly showed a good correspondence to the components of the lacquer sap.

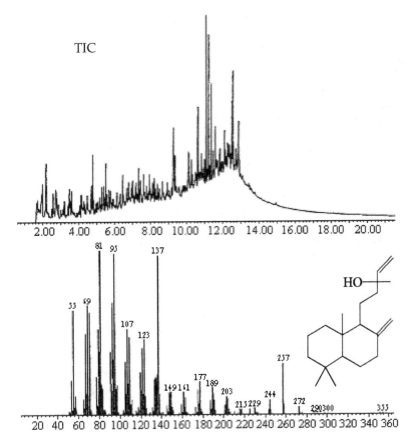

Fig. 4-21. Pyrolysis data of Baroque and Rococo lacquer films

5. Conclusion

This chapter describes the chemical properties of lacquer saps and films that were analyzed by pyrolysis-gas chromatography/mass spectrometry measurement. It was revealed that the advanced Py-GC/MS analytical method is useful for identifying and evaluating the lacquer components and the origin of lacquer species. Py-GC/MS is a well-known method applied to various areas. Due to its ease of control, speed of analysis, and good reproducibility, the Py-GC/MS method not only can be applied to lacquer films, organic coatings, and other materials that cannot be dissolved in solvents, but also to discriminate between lacquer and other resins for the conservation or restoration of lacquerware.

6. References

[1] J. Gan. Lacquer Chemistry, Chinese Academy of Science, 1984.
[2] O. Vogl, J. Bartus, M. Qin, J. Mitchell. Progress in Pacific Polymer Science 3, Spring-Verlag,Berlin, 1994, p. 423.

[3] O. Vogl. Oriental lacquer, poison ivy, and drying oils. *J. Polym. Sci. A: Polym. Chem.* 38, 2000, 4327-4335.

[4] E. J. Kidder. Ancient Peoples and Places, Japan, Thames & Hudson, 1959.

[5] Y. Kuraku. Urushi, N. S. Brommelle and P. Smith (eds.), Getty Conservation Institute, California, 45, 1988.

[6] Y. M. Du. Urushi, N. S. Brommelle and P. Smith (eds.), Getty Conservation Institute, California, 194, 1988.

[7] J. Hu. Conservation and Restoration of Cultural Property: Conservation of Far Eastern Objects, Tokyo National Res. Inst. Of Cultural Property, Tokyo, 89-112, 1980.

[8] A. Huttermann, C. Mai, and A. Kharazipour. Modification of lignin for the production of new compounded materials. *Appl. Microbiol. Biotechnol.*, 55, 2001, 387-394.

[9] http://inoues.net/ruins/torihama.html.

[10] http://www.daily-tohoku.co.jp/kikaku/tyouki-kikaku/jomon/jomon-22.htm.

[11] J. Kumanotani. Urushi (Oriental lacquer) − a natural aesthetic durable and future-promising coating. *Prog. Org. Coat.*, 26, 1995, 163-195.

[12] T. Miyakoshi, K. Nagase, T. Yoshida. Progress of Lacquer Chemistry, IPC Publisher, Tokyo, Japan, 1999, p. 43.

[13] T. Terada, K. Oda, H. Oyabu, T. Asami. Lacquer - the Science and Practice, Rikou Publisher, Tokyo, Japan, 1999, p. 25.

[14] Y. Yamauchi, R. Oshima, J. Kumanotani. Configuration of the olefinic bonds in the heteroolefinic side-chains of Japanese lacquer urushiol: separation and identification of components of dimethylurushiol by means of reductive ozonolysis and high- performance liquid chromatography. *J. Chromatogr.*, 243, 1982, 71-84.

[15] B. Reinhammar, Y. Oda. Spectroscopic and catalytic properties of *Rhus vernicifera* laccase depleted in type 2 copper. *J. Inorg. Biochem.*, 11, 1979, 115-127.

[16] J. Kumanotani. Urushi − cultural relations. *Kagakutokougyo*, 36, 1983, 151-154.

[17] S. Harigaya, T. Honda, R. Lu, T. Miyakoshi, and C. Chen. Enzymatic dehydrogenative polymerization of urushiols in fresh exudates from the lacquer tree, *Rhus vernicifera* DC. *J. Agric. Food Chem.*, 55, 2007, 2201-2208.

[18] R. Lu, S. Harigaya, T. Ishimura, K. Nagase, T. Miyakoshi. Development of a fast drying lacquer based on raw lacquer sap. *Prog Org. Coat.*, 51, 2004, 238-243.

[19] R. Lu, T. Ishimura, K. Tsutida, T. Honda, T. Miyakoshi. Development of a fast drying hybrid lacquer in a low relative-humidity environment based on *Kurome* lacquer sap. *J. Appl. Polym. Sci.*, 98, 2005, 1055-1061.

[20] R. Lu, D. Kanamori, and T. Miyakoshi. Characterization of thitsiol dimer structure from *Melanorrhoea usitata* with laccase catalyst by NMR spectroscopy. *Inter. J. Polym. Anal. Character.*, 16, 2011, 86-94.

[21] M. Herreraa, G. Matuschek. Fast identification of polymer additives by pyrolysis-gas chromatography/mass spectrometry. *J. Anal. Appl. Pyrolysis*, 70, 2003, 35-42.

[22] T. Honda, R. Lu, N. Kitano, Y. Kamiya, T. Miyakoshi. Applied analysis and identification of ancient lacquer based on pyrolysis-gas chromatography/mass spectrometry. *J. Appl. Polym. Sci.*, 118, 2010, 897-901.

[23] R. Lu, Y. Kamiya, Y. Wan, T. Honda, T. Miyakoshi. Synthesis of *Rhus succedanea* lacquer film and analysis by pyrolysis-gas chromatography/mass spectrometry. *J. Anal. Appl. Pyrolysis*, 78, 2007, 117-124.

[24] Y. Wan, R. Lu, Y. Du, T. Honda, T. Miyakoshi. Does Donglan lacquer tree belong to *Rhus vernicifera* species? *Inter. J. Bio. Macromol.*, 41, 2007, 497-503.

[25] K. Urakami, A. Higashi, K. Umemoto, M. Godo, C. Watanabe, K. Hashimoto. Compositional analysis of copoly(DL-lactic/glycolic acid) (PLGA) by pyrolysis-gas chromatography/mass spectrometry combined with one-step thermally assisted hydrolysis and methylation in the presence of tetramethylammonium hydroxide. *Chem. Pharm. Bull.*, 49, 2001, 203-205.

[26] M. Fukushima, M. Yamamoto, T. Komai, K. Yamamoto. Studies of structural alterations of humic acids from conifer bark residue during composting by pyrolysis-gas chromatography/mass spectrometry using tetramethylammonium hydroxide (TMAH-py-GC/MS). *J. Anal. Appl. Pyrolysis*, 86, 2009, 200-206.

[27] X. Chai, T. Shimaoka, Q. Guo, Y. Zhao. Characterization of humic and fulvic acids extracted from landfill by elemental composition, 13C CP/MAS NMR and TMAH-Py-GC/MS. *Waste Management*, 28, 2008, 896-903.

[28] R. Widyorini, T. Higashihara, J. Xu, T. Watanabe, S. Kawai. Self-bonding characteristics of binderless kenaf core composites. *Wood Sci. Tech.*, 39, 2005, 651-662.

[29] R. Martine, G. Nicolas, C. Cecile, R. Christian. New analytical methodology for investigating both volatile and nonvolatile constituents from archaeological ceramic vessels, Abstracts of Papers, 225th ACS National Meeting, New Orleans, LA, United States, March 23-27, 2003, GEOC-107.

[30] H. Araki, N. Tatarazako, K. Kishi, K. Kuroda. Evaluation of bioaccumulation potential of 3,4,5-trichloroguaiacol in a zooplankton (*Daphnia magna*) by pyrolysis-GC/MS in the presence of tetramethylammonium hydroxide (TMAH). *J. Anal. Appl. Pyrolysis*, 55, 2000, 69-80.

[31] T. Arii. Complex thermal analysis: evolved gas analysis-mass spectrometry EGA-MS. *Rigaku Janaru*, 39, 2008, 17-25.

[32] T. Arii. Evolved gas analysis-mass spectrometry (EGA-MS) using skimmer interface system equipped with pressure control function. *J. Mass Spectro. Soc. Jap.*, 53, 2005, 211-216.

[33] T. Tsugoshi, M. Furukawa, M. Ohashi, Y. Iida. Comparison of capillary and skimmer interfaces in evolved gas analysis-mass spectrometry (EGA-MS) with regard to impurities in ceramic raw materials. *J. Therm. Anal. Calorimetry*, 64, 2001, 1127-1132.

[34] G. Matuschek, A. Kettrup, A. Prior. EGA/MS investigations on the thermal degradation of diammoniumhexachloroplatinate. *J. Therm. Anal. Calorimetry*, 56, 1999, 471-477.

[35] J. Lee, G. Chang, J. Kwag, M. Rhee, A. Buglass, G. Lee. Fast analysis of nicotine in tobacco using double-shot pyrolysis-gas chromatography-mass spectrometry. *J. Agric. Food Chem.*, 55, 2007, 1097-1102.

[36] K. Quenea, S. Derenne, F. Gonzalez-Vila, J. Gonzalez-Perez, A. Mariotti, C. Largeau. Double-shot pyrolysis of the non-hydrolysable organic fraction isolated from a sandy forest soil (Landes de Gascogne, South-West France). *J. Anal. Appl. Pyrolysis*, 76, 2006, 271-279.

[37] J. Lee, C. Lee, J. Kwag, A. Buglass, G. Lee. Determination of optimum conditions for the analysis of volatile components in pine needles by double-shot pyrolysis-gas chromatography-mass spectrometry. *J. Chromatogr. A*, 1089, 2005, 227-234.

[38] R. Lu, Y. Kamiya, and T. Miyakoshi. Applied analysis of lacquer films based on pyrolysis-gas chromatography/mass spectrometry. *Talanta*, 70, 2006, 370-376.

[39] N. Niimura, T. Miyakoshi, J. Onodera, and T. Higuchi. Structural studies and polymerization mechanisms of synthesized lacquer films using two-stage pyrolysis-gas chromatography/mass spectrometry. *Inter. J. Poly. Anal. Charact.*, 4, 1998, 309-322.

[40] N. Niimura, T. Miyakoshi, J. Onodera, and T. Higuchi. Identification of ancient lacquer film using two-stage pyrolysis-gas chromatography/mass spectrometry. *Archaeometry*, 41 1999, 137-149.

[41] R. Lu, Y. Kamiya, and T. Miyakoshi. Preparation and characterization of *Melanorrhoea usitata* lacquer film based on pyrolysis-gas chromatography/mass spectrometry. *J. Anal. Appl. Pyrolysis*, 78, 2007, 172-179.

[42] Y. Kamiya, Y. Niimura, and T. Miyakoshi. Evaluation of synthesized lacquer dilms using pyrolysis-gas chromatography/mass spectrometry. *Bull. Chem. Soc. Jpn.*, 73, 2000, 2621-2626.

[43] R. Lu, Y. Kamiya, and T. Miyakoshi. Characterization of lipid components of *Melanorrhoea usitata* lacquer sap. Talanta, 71, 2007, 1536-1540.

[44] R. Lu, X. Ma, Y. Kamiya, T. Honda, Y. Kamiya, A. Okamoto, and T. Miyakoshi. Identification of Ryukyu lacquerware by pyrolysis-gas chromatography/mass spectrometry. *J. Anal. Appl. Pyrolysis*, 80, 2007, 101-110.

[45] N. Niimura, and T. Miyakoshi. Characterization of natural resin films and identification of ancient coating. *J. Mass Spectrom. Soc. Jpn.*, 51, 2003, 439-457.

Pyrolysis-Gas Chromatography to Evaluate the Organic Matter Quality of Different Degraded Soil Ecosystems

Cristina Macci, Serena Doni, Eleonora Peruzzi,
Brunello Ceccanti and Grazia Masciandaro
Consiglio Nazionale delle Ricerche (CNR),
Istituto per lo Studio degli Ecosistemi (ISE), Pisa,
Italy

1. Introduction

Soil systems are exposed to a variety of environmental stresses of natural and anthropogenic origin, which can potentially affect soil functioning. For this reason, there is growing recognition for the need to develop sensitive indicators of soil quality that reflect the effects of land management on soil and assist land managers in promoting long-term sustainability of terrestrial ecosystems (Bandick & Dick, 1999). Soil organic matter (SOM) providing energy, substrates, and biological diversity necessary to sustain numerous soil functions, has been considered one of the most important soil properties that contributes to soil quality and fertility (Doran & Parkin, 1994; Reeves, 1997).

SOM comprises a range of humic substances (HS) and non humic substances (NHS). Humic substances make up a significant portion of the total organic carbon (TOC) and nitrogen (TN) in soil. They consist of complex polymeric organic compounds of high molecular weights, which are more resistant to decomposition than the NHS. Moreover, humic substances are widely recognized as an important fraction of soil organic matter because they have several fundamental functions: regulation of nutrient availability, linkage with mineral particles and immobilization of toxic compounds (Ceccanti & Garcia, 1994).

On the other hand, NHS contribute to soil ecosystem functionality by providing metabolically labile organic C and N sources, such as low molecular weight aliphatic and aromatic acids, carbohydrates, aminoacids, and their polymeric derivatives such as polypeptides, proteins, polysaccharides, and waxes (Schnitzer, 1991).

In order to understand the temporal dynamics of SOM in soil ecosystems, it is therefore vital to characterize soil organic carbon quantity and quality. Various analytical methodologies, such as infrared (See et al., 2005), ultraviolet–visible (Alberts et al., 2004), nuclear magnetic resonance spectroscopy (Gondar et al., 2005), potentiometric titration, pyrolysis, electrophoresis, acid/alkali fractionation, have been used to study SOM (Ceccanti et al., 2007). These techniques have been used either alone or in combination with traditional fractionation and purification methods of acid/alkali soluble HS. However, these methodologies may not always reflect native SOM, due to inevitable analytical manipulations and also, sometimes, to a subjective interpretation of the results.

Among these techniques, pyrolysis-gas chromatography (Py-GC) is a technique capable of analysing native SOM, both in bulk soil (without manipulation), and in sodium pyrophosphate extract, considered the favoured solution to extract humic organic matter from soil (García et al., 1992; Masciandaro & Ceccanti, 1999; Clapp & Hayes, 1999). However, some disadvantages exist, since this technique provides information on small molecules but not the overall macromolecular structure (Grandy & Neff, 2008). Nevertheless, Py-GC is a simple and rapid technique which has been used successfully to make a qualitative study of the chemico-structural characteristics of organic matter in soils, sewage sludges (Peruzzi et al., 2011), sediments (Faure et al., 2002), fresh and composted wastes (Garcia et al., 1992; Dignac et al., 2005).

Many papers has been published in the last years on the application of Py-GC to soil analyses (Campo et al., 2011; Sobeih et al., 2008). Nierop et al. (2001) investigated the differences in the chemical composition of SOM within one soil serie from three differently managed fields using a combination of Py-GC/MS and thermally assisted hydrolysis and methylation. The results of this study showed that SOM composition is hardly affected by organic farming compared to conventional management i.e. high tillage intensity and intensive fertilization. Similarly, Marinari et al. (2007a, 2007b) used pyrolytic indices as indicators of SOM quality under organic and conventional management. In these studies has been confirmed that mineral fertilization, showing a relatively large amounts of aliphatic pyrolytic products, contributed to increase the mineralization process.

The Py-GC resulted a promising technique in the assessment of change in soil organic matter composition, even after the application of different types of mulching materials (Ceccanti et al., 2007). In the study of Ceccanti et al. (2007) compost enriched the soil with available organic nutrients, rapidly shown by an increase of pyrolytic aliphatic compounds, while straw application revealed a more stable soil organic matter characterized by a high pyrolytic humification index.

In this chapter the Py-GC technique will be proposed as useful tool to evaluate, on the basis of the chemical structural changes of organic matter, the resistance of degraded soils to different external impacts and rehabilitation practices.

The results obtained from the study of several sites, characterized by local variation of anthropic and natural pressures, will be presented. The investigation has been carried out in the framework of the European project "Indicators and thresholds for desertification, soil quality, and remediation" (INDEX). In this project 12 sites were selected across Europe in order to ensure a broad range of different conditions (natural soil and agro-ecosystems).

The parameters determined by pyrolysis-gas chromatography have been selected as promising indicators of soil degradation and rehabilitation. These parameters were mainly classified as indicator of "State" according to DPSIR (Driving Forces-Pressure-State-Impact-Response) framework of Thematic Strategy for Soil Protection (European Commission, 2006) since making a fingerprint of soil organic matter, they represent the evolution state of organic matter in different climatic and management situation.

2. Methodologies

2.1 Pyrolysis-gas chromatography

Py-GC technique is based on a rapid decomposition of organic matter under a controlled high flash of temperature. A gas chromatograph is used for the separation and quantification of pyrolytic fragments (pyro-chromatogram) originating from organic matter decomposition. 50 micrograms of a representative soil sample, air-dried and ground <100 mesh (bulk soil) or few ml of soil sodium pyrophosphate extract (0.1M) was introduced into pyrolysis quarz microtubes in a CDS Pyroprobe 190.

The textural characteristics of soil can affect the reliability of the Py-GC technique, in particular in degraded soils, for this reason the method can be adapted in order to remove the inert part (for example sand and gross silt) which is not involved in the linkage with the humic substances, in other words, the soil organic matter has to be concentrated. Moreover, for highly organic soils (organic matter > 5%), where a great part of organic matter is linked to quick microbial metabolism, the extraction of humic matter (e.g. sodium pyrophosphate) give more reliable information on native soil organic matter. The Py-GC instrument consisted of a platinum coil probe and a quartz sample holder. Pyrolysis was carried out at 800 °C for 10 s, with a heating rate of 10 °C ms^{-1} (nominal conditions). The probe was coupled directly to a Carlo Erba 6000 gas chromatograph with a flame ionization detector (FID). Chromatographic conditions were as follows: a 3 m x 6 mm, 80/100 mesh, SA 1422 (Supelco inc) POROPAK Q packed column, the temperature program was 60 °C, increasing to 240 °C by 8 °C min^{-1}.

Pyrograms were interpreted by quantification of seven peaks corresponding to the major volatile pyrolytic fragments (Ceccanti et al., 1986): acetic acid (K), acetonitrile (E$_1$), benzene (B), toluene (E$_3$), pyrrole (O), furfural (N), and phenol (Y). The retention time (minutes) of the mentioned seven peaks (as standard) are the following: E1, 20.2, K, 23.1, B, 27.8, O, 30.9, E3, 34.7, N, 38.9, Y, 55.6 (Scheme 1). Acetonitrile (E$_1$) is derived from the pyrolysis of aminoacids, proteins, and microbial cells. Furfural (N), is mostly derived from carbohydates, ligno-cellulosic materials, proteins and other aliphatic organic compounds (Bracewell & Robertson, 1984), indicating the presence of rapidly metabolisable organic substances. Acetic acid (K) is preferentially derived from pyrolysis of lipids, fats, and waxes, cellulose, carbohydrates (Bracewell & Robertson, 1984) and represents relatively less-degraded ligno-cellulose material (Sollins et al., 1996; Buurman et al., 2007). Phenol (Y) is derived from amino acids, tannins and fresh or condensed (humic) lignocellulosic structures (van Bergen et al., 1998; Lobe et al., 2002). Benzene (B) and toluene (E$_3$) are basically derived from condensed aromatic structures of stable (humified) organic matter, particularly for benzene, since toluene must come from rings with aliphatic chains, albeit short. Pyrrole (O) is derived from nitrogenated compounds such as nucleic acids, proteins, microbial cells and condensed humic structures (Bracewell & Robertson, 1984).

Peak areas were normalized, so that the area under each peak referred to the percentage of the total of the selected seven peaks (relative abundances).

The alphabetic code used was conventional and has already been used in previous papers on natural soils (Ceccanti et al., 1986; Alcaniz et al., 1984). Peak purity of the most important fragments had previously been checked by coupling the same chromatographic system to a mass-detector under the same operating conditions.

Ratios of the relative abundances of the peaks were determined (Ceccanti et al., 1986, 2007) as follows:

N/O: mineralization index of labile soil organic matter. This index expresses the ratio between furfural (N), which is the pyrolytic product arising from polysaccharides, and pyrrole (O), which derives from nitrogenated compounds, humified organic matter, and microbial cells. The higher the ratio, the lower the mineralization of the organic matter, that means a high concentration of polysaccharides may be present.

O/Y: mineralization index of more-stable OM. The higher the ratio, the higher the mineralization of the organic matter, because pyrrole (O) derives from nitrogenated compounds and microbial cells and phenol (Y) mostly derives from polyphenolic compounds.

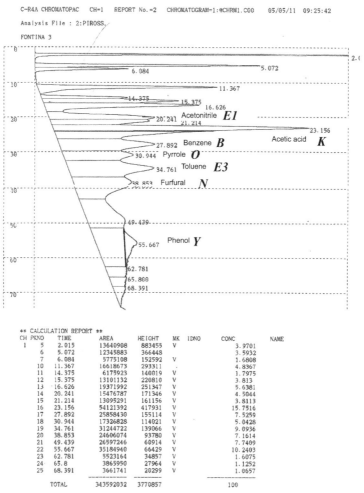

Scheme 1. Pyrogram of sample with time retention and area of each fragment: Acetonitrile (E1), Acetic acid (K), Benzene (B), Pyrrole (O), Toluene (E3), Furfural (N), Phenol (Y).

B/E_3: humification index. The higher the ratio, the higher the condensation of the organic matter, because benzene (B) derives mostly from the pyrolytic degradation of condensed aromatic structures, while toluene (E_3) comes from aromatic uncondensed rings with aliphatic chains.

AL/AR (Aliphatic/Aromatic compounds): index of "energetic reservoir". It expresses the ratio between the sum of aliphatic products (acetic acid K, furfural N, and acetonitrile E_1) and the sum of aromatic compounds (benzene B, toluene E_3, pyrrole O and phenol Y). This index could indicate an "energetic reservoir" especially in extremely poor soils since it evaluates the labile and stable parts of HS.

In addition, a numeric index of similarity (Sij) between the relative abundances (I) of the homologous peaks (k) in two pyro chromatograms (i and j), was calculated using the following expression: $Sij = (\sum(Ii/Ij)k)/n$ with $Ii < Ij$, and n is the number of peaks (Ceccanti et al., 1986).

The index of similarity Sij compares a pair of pyrograms without discriminating peaks. The index varies in the range of 0–1: the higher the value, the greater the similarity. However, three conventional levels, very high (>0.85), high (0.75-0.85), middle (0.70–0.75), and low (0.60–0.70) have been suggested for the characterization of heterogeneous materials such as SOM (Ceccanti et al., 1986) and compost (Ceccanti & Garcia, 1994; Ceccanti et al., 2007).

2.2 Humic substances extraction

Soil organic matter comprises a range of humified, non-humified, and biologically active compounds, including readily decomposable materials, plant litter and roots, and dead and living organisms. The humified compounds, representing the most microbially recalcitrant and thus stable reservoir of organic carbon in soil may be more appropriate than the labile organic matter compounds in the definition of "soil quality", they can, in fact, be considered a sort of recorder of the soil's history (Masciandaro & Ceccanti, 1999; Piccolo et al., 1996).

Humic substances based on solubility under acidic or alkaline conditions can be fractionated as following: humin, the insoluble fraction of humic substances, humic acid (HA), the fraction soluble under alkaline conditions but not acidic conditions (generally pH <2), and fulvic acid (FA), the fraction soluble under all pH conditions.

Due to the low solubility and complex chemical structure of these compounds, soil extraction and purification are essential steps to study humic substances (Ceccanti et al., 1978). Sodium pyrophosphate is the favoured solution used to extract organic matter from soil, and, neutral conditions permit to extract the greater part of humic carbon, maintaining the functional properties of soil (Masciandaro & Ceccanti,1999).

The extraction of humic substances from the soils were carried out by mechanical shaking of the soil samples with 0.1 M $Na_4P_2O_7$ solution at pH 7.1 for 24 h at 37 °C in a soil/solution ratio of 1/10. The extracts were centrifuged at 24 000 x g for 20 min and filtered on a 0.22 µm Millipore membrane. Identical volumes of these extracts were dialysed in Visking tubes with distilled water. After dialysis, extracts were concentrated by ultrafiltration (AMICON PM10 cut-off membrane) and brought to the volume that extracts had prior to dialysis (Nannipieri et al., 1983; Masciandaro et al., 1997).

3. Case studies

Five case studies have been selected in order to validate the pyrolysis gas-chromatography methodology as useful tool to evaluate chemico-structural changes of soil organic matter: i) two natural soil ecosystems (Catena) with different density of natural plant cover (natural soil), one located in the south of Spain and the other in the south of Italy, ii) two sites under organic (biological) and mineral (conventional) management, one in Cyprus Island and one in the centre of Italy, and iii) a degraded soil undergone to rehabilitation practice with different amount of organic matter (restored soils).

Each soil sample has been collected at 0-15 cm depth and the pyrolysis gas-cromatography has been carried out both in the bulk soil, and in sodium pyrophosphate extract of each sample.

3.1 Natural soil

3.1.1 Spanish catena

A site in Santomera area (Murcia region, Spain) characterized by three different gradual degradation states, related to different natural plant cover establishment and soil slope, has been sampled (sandy loam, USDA classification): i) a natural soil with a 50-60% of vegetation cover of *Pinus halepensis* (high plant density, **Forest**) and a soil slope of 5%, ii) a partially degraded soil with a 20-30% of vegetation cover (autochthonous xerophytic shrub) (low plant density, **Shrub**) and slope of 5%, and iii) a bare soil with only a 5-10% of vegetation cover (scant plant density, **Bare**) and slope of 10%.

The main difference among the soil samples was presented by the percentage of toluene (E_3), the significant increase along the catena ecosystem (Bare > Shrub > Forest) (Table 1), indicated the prevalence of organic matter less-humified in Bare and Shrub with respect to Forest soil, since E_3 comes from aromatic uncondensed rings with aliphatic chains.

In agreement with this result, lower values of benzene (B) and phenol (Y), which derived principally from humified stable organic matter, were found in Bare soil with respect to Forest and Shrub soil, suggesting the reduction of condensed aromatic structure of stable humified organic matter (Masciandaro & Ceccanti, 1999) in consequence of a fast metabolism, involving probably the native humic matter.

bulk soil	E_1%	K%	B%	O%	E_3%	N%	Y%
Forest	11,7	11,7	18,0	22,9	15,4	17,6	2,60
Shrub	11,5	11,3	18,5	20,3	18,5	17,1	2,72
Bare	13,2	10,0	17,4	21,1	19,8	16,8	1,66

Table 1. Relative abundances (%) of main pyrolytic peaks (acetonitrile E_1, acetic acid K, benzene B, pyrrole O, toluene E_3, furfural N, phenol Y) in bulk soils.

Soil differentiation was better shown by the pyrolytic indexes of humification (B/E_3) and mineralization (O/Y and N/O). The humification index B/E_3 which represents the structural condition of a 'condensed aromatic nucleus' of humic substances, increased with the increase of the intensity of vegetation (Forest > Shrub > Bare) (Figure 1), indicating the importance of plant cover in the storage of more stable humus and in the reduction of degradation process (Vancampenhout et al., 2010).

In Bare soil, the scant vegetation seemed unable to protect the soil organic matter from mineralization, as confirmed by the higher value of O/Y (Figure 1). However, the N/O mineralization index, that gave an evidence of the existence of an easily mineralizable fraction of organic matter, did not show many differences among the soils (Figure 1), suggesting that the labile organic matter was less influenced.

Percentages of stable benzene-derived aromatic (AR = B + E_3 + O + Y) and mineralizable aliphatic (AL = E_1 + K + N) structures were very similar in all the three soils (Figure 1). This would indicate that, despite the highest mineralization occurred in Bare soil, the catena ecosystem was self-regulating and still preserved a significative amount of microbial resistant humic substances (Ceccanti & Masciandaro, 2003).

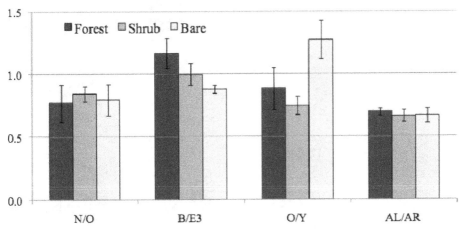

Fig. 1. Indices of mineralization (furfural/pyrrole N/O, pyrrole/phenol O/Y), humification (benzene/toluene B/E_3), and aliphatic/aromatic (AL/AR) ratios (AL=E_1+K+N, AR=B+O+E_3+Y) (mean of three replicates ± standard deviation) in bulk soils The values of O/Y have been divided by 10.

This sort of natural balance in the Catena ecosystem, was confirmed by the high indexes of similarity observed among the three soil samples and, in particular, between Forest and Shrub (Table 2).

bulk soil	Shrub	Bare
Forest	0,938	0,857
Shrub		0,883

Table 2. Pyrolitic index of similarity in the bulk soils.

In the soil extract, the chemical and structural information is amplified by the fact that unlike the bulk soil, the humic fraction is more concentrated and pure and free from interfering substances. In this case no great differences were generally observed among the soils, both in the content of pyrolytic fragments (Table 3) and calculated indices (Figure 2). However, the higher value of N/O found in Bare with respect to Forest soil, suggested that the mineralization of organic matter, hypothesized on the basis of mineralization index O/Y of bulk soil, involved the labile fraction of humic matter. The high similarity indexes observed among the bulk soils (Table 2) were maintained also among the humic extracts (Table 4), underlining a better homogeneity considering the humic part of the organic matter.

Soil extract	$E_1\%$	K%	B%	O%	$E_3\%$	N%	Y%
Forest	8,26	8,53	16,6	19,7	27,2	13,4	6,31
Shrub	8,43	7,48	14,9	18,8	26,3	15,1	9,31
Bare	6,98	10,0	18,2	17,1	27,4	15,4	4,86

Table 3. Relative abundances (%) of main pyrolytic peaks (acetonitrile E_1, acetic acid K, benzene B, pyrrole O, toluene E_3, furfural N, phenol Y) in soil extracts.

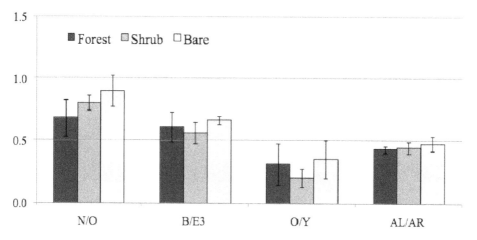

Fig. 2. Indices of mineralization (furfural/pyrrole N/O, pyrrole/phenol O/Y), humification (benzene/toluene B/E_3), and aliphatic/aromatic (AL/AR) ratios (AL=E_1+K+N, AR=B+O+E_3+Y) (mean of three replicates ± standard deviation) in soil extracts. The values of O/Y have been divided by 10.

soil extract	Shrub	Bare
Forest	0,872	0,888
Shrub		0,820

Table 4. Pyrolitic index of similarity in the soil extracts.

3.1.2 Italian catena

A natural catena similar to the Spanish one, were studied in the South of Italy (Basilicata region) (sandy, clay, loam, USDA classification).

Three different gradual degradation states, related to different natural plant cover establishment and soil slope, has been sampled: i) a natural soil with a 60-70% of vegetation cover of *Quercus pubescens* Willd (high plant density, **Forest**) and a soil slope of 15%, ii) a partially degraded soil with a 20-30% of vegetation cover, consisting in autochthonous xerophytic shrub and grass (low plant density, **Shrub**) and slope of 6%, and iii) an eroded soil without plant cover (no plant, **Bare**) and slope of 20%.

In this Italian natural Catena, differently from the previous, the main difference among the three soils concerned the pyrrole (O) and particularly phenol (Y) fragments (Table 5). An increase of phenol in Bare suggested a slower soil metabolism that slightly affected the dinamic of stable humic matter, probably due to the lack of vegetation cover. The simultaneous increase in pyrrole (O) and phenol (Y) fragments in each soil, indicated the same level of mineralization processes of more stable soil organic matter in the three different degraded soils.

However, as already found for Spanish catena, there was a decrease in humification index with the decrease of vegetal cover (Figure 3). This trend, together with the lower value of AL/AR and N/O in Bare soil, suggested that a great mineralization process interesting the labile part of the organic matter affected this environment.

A lower similarity was, in fact, found among the soils and in particular between Forest and Bare (Table 6) soils.

bulk soil	E_1%	K%	B%	O%	E_3%	N%	Y%
Forest	16,0	9,53	16,1	10,6	24,5	10,0	14,2
Shrub	11,9	11,2	14,6	11,2	23,0	11,8	16,2
Bare	8,75	11,2	11,0	14,5	24,6	8,09	21,9

Table 5. Relative abundances (%) of main pyrolytic peaks (acetonitrile E_1, acetic acid K, benzene B, pyrrole O, toluene E_3, furfural N, phenol Y) in bulk soils.

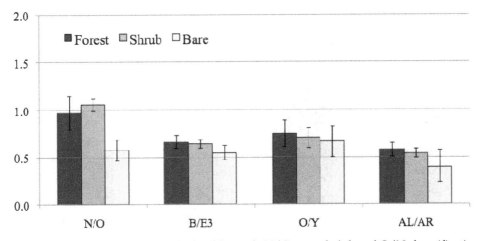

Fig. 3. Indices of mineralization (furfural/pyrrole N/O, pyrrole/phenol O/Y), humification (benzene/toluene B/E_3), and aliphatic/aromatic (AL/AR) ratios (AL=E_1+K+N, AR=B+O+E_3+Y) (mean of three replicates ± standard deviation) in bulk soils.

bulksoil	Shrub	Bare
Forest	0,845	0,752
Shrub		0,801

Table 6. Pyrolitic index of similarity in the bulk soils

Soil extract	E1%	K%	B%	O%	E3%	N%	Y%
Forest	8,94	11,4	23,4	9,90	20,0	12,3	13,9
Shrub	13,8	13,9	19,2	12,9	14,2	14,5	11,5
Bare	6,39	16,1	25,7	12,0	22,0	9,55	8,32

Table 7. Relative abundances (%) of main pyrolytic peaks (acetonitrile E_1, acetic acid K, benzene B, pyrrole O, toluene E_3, furfural N, phenol Y) in soil extracts.

In the humic extract, the trend of the mineralization indices O/Y and N/O, highlighted the hard mineralization process that affect bare soil, followed by shrub soil (Figure 4, Table 7).

In this Catena ecosystem, differently from the previous, the scant vegetation and probably other environmental factors such as greater soil slope and Mediterranean climate, with dry hot summer and cold winters with strong storm, make the bare soils more susceptible to erosion processes.

However, the high values of B/E_3 detected in each soil, suggested that this ecosystem is able to preserve the more condensed humic matter.

As already observed for the bulk soils, medium-high similarity indices were found among the humic extracts of the different situations (Table 8). Therefore, both Italian and Spanish Catena, were more homogeneous considering the humic part of the organic matter.

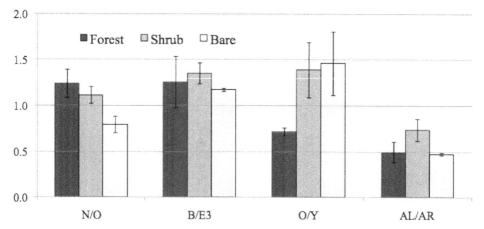

Fig. 4. Indices of mineralization (furfural/pyrrole N/O, pyrrole/phenol O/Y), humification (benzene/toluene B/E_3), and aliphatic/aromatic (AL/AR) ratios ($AL=E_1+K+N$, $AR=B+O+E_3+Y$) (mean of three replicates ± standard deviation) in soil extracts.

Soil extract	Shrub	Bare
Forest	0,778	0,778
Shrub		0,718

Table 8. Pyrolitic index of similarity in the soil extracts

3.2 Organic and mineral management

3.2.1 Site in Cyprus Island

Two adjacent fields in Cyprus Island, where one was managed according to biological, and the other according to conventional farming methods, were studied (sandy clay loam soil, USDA classification). In the biological agriculture, the fields were fertilized with 10 t ha^{-1} y^{-1} of commercial manure (in pellets), while in the conventional treatment mineral fertiliser was added at a total rate of 300 kg ha^{-1}.

A clear difference between biological and conventional management was evident in Py-GC fragments analysis (Table 9): a higher abundance of benzene (B) and phenol (Y) characterized the biological treatment, thus indicating the establishment of a humified stable organic matter nucleus. On the other hand, the greater value of toluene (E$_3$) and pyrrole (O) found in conventional treatment highlighted the presence of humified organic matter, especially that in less condensed and pseudo-stable form (Table 9). In addition, the high presence of pyrrole (O) in conventional management suggested the triggering of mineralization processes. In fact, the higher value of O/Y and the lower value of N/O in conventional treatment (Figure 5), highlighted the higher level of mineralization of less mineralizable and easily degradable organic matter, respectively, occurred in this soil. Instead, the impressive degree of index of humification B/E$_3$ found in biological treatment suggested the incorporation of organic matter added to soil addressed the metabolic processes towards the organic matter humification.

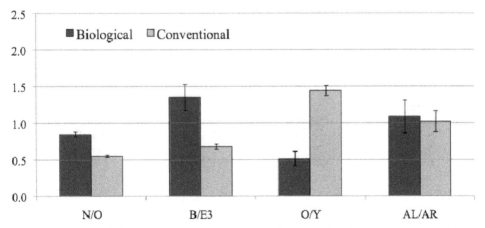

Fig. 5. Indices of mineralization (furfural/pyrrole N/O, pyrrole/phenol O/Y), humification (benzene/toluene B/E$_3$), and aliphatic/aromatic (AL/AR) ratios (AL=E$_1$+K+N, AR=B+O+E$_3$+Y) (mean of three replicates ± standard deviation) in bulk soils.

bulk soil	E_1%	K%	B%	O%	E_3%	N%	Y%
Biological	26,1	20,3	16,8	6,23	12,8	5,29	12,4
Conventional	29,1	15,5	13,0	10,4	19,0	5,75	7,21

Table 9. Relative abundances (%) of main pyrolytic peaks (acetonitrile E_1, acetic acid K, benzene B, pyrrole O, toluene E_3, furfural N, phenol Y) in bulk soils.

The results obtained from the analysis of humic extracts, discriminated more the different influence of the two management practices on soil organic matter quality (Table 10). The effect of compost addition on soil under biological treatments was, in fact, highlighted by the higher level of acetic acid (K) and pyrrole (O) relative abundances with respect to the conventional one, as also found by Ceccanti et al. (2007). The values of these compounds suggested the presence of fresh organic matter and the beginning of a humification process mediated by microbial activity (Peruzzi et al., 2011). The higher relative abundance of toluene (E_3) in conventional site was in agreement with the results of bulk soil pyrolytic fragments, thus showing the presence of less stable humic fraction. The index of mineralization N/O (Figure 6) showed that, in biological treatment, mineralization process of easily degradable organic matter occurred. However, the highest index of humification in biological site, as shown by pyrolytic indices in the bulk soil, suggested that the organic matter added with compost has became part of humic pool of soil (Garcia et al.,1992).

The differences occurred in management practises could be resumed by similarity indices, which clearly expressed the difference between the two treatments both in bulk soil and in pyrophosphate extract (S_{soil}: 0.743, $S_{extract}$: 0.774).

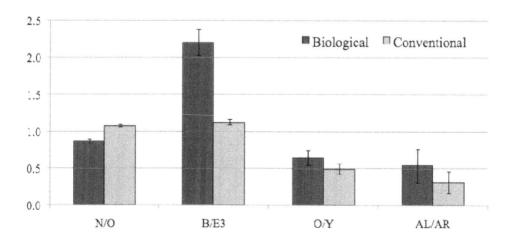

Fig. 6. Indices of mineralization (furfural/pyrrole N/O, pyrrole/phenol O/Y), humification (benzene/toluene B/E_3), and aliphatic/aromatic (AL/AR) ratios (AL=E_1+K+N, AR=B+O+E_3+Y) (mean of three replicates ± standard deviation) in soil extracts.

Soil extract	E_1%	K%	B%	O%	E_3%	N%	Y%
Biological	4,20	17,88	18,9	14,6	8,63	12,6	23,2
Conventional	4,49	6,95	21,5	11,7	19,2	12,4	23,7

Table 10. Relative abundances (%) of main pyrolytic peaks (acetonitrile E_1, acetic acid K, benzene B, pyrrole O, toluene E_3, furfural N, phenol Y) in soil extracts.

3.2.2 Site in Italy

Two adjacent fields in Central Italy (Tuscany region), one managed according to biological, and the other according to conventional farming methods, were also studied. The soil of both fields was classified as sandy clay loam according to the USDA classification. In the biological agricultural soil (biological procedure for five years), green manure (1,5 t ha^{-1} y^{-1}) was applied to the fields, while in the conventional agricultural soil, the fields received mineral fertilizer (0,2 t ha^{-1} y^{-1} ammonium nitrate).

The Italian situation showed fewer differences both in soil and extract, considering the two management practices, with respect to the Cyprus case (Similarity indices S_{soil}: 0.873, $S_{extract}$: 0.870). However, a higher level of benzene (B) and pyrrole (O) characterized biological treatment (Table 11). This result suggested that this kind of soil management promoted: i) the establishment of more humified organic matter, as also suggested by the trend of humification index B/E3, and ii) the stimulation of microbial metabolism driving the mineralization processes of the available organic matter, as suggested by the trend of O/Y and N/O indices (Figure 7). In other words, the biological management stimulated more the process of organic matter turnover, with respect to the conventional one, which on the contrary, started the mineralization of the humified organic matter, as suggested by the highest value of O/Y in the humic extract (Figure 8, Table 12). However, the trend of humification index B/E_3 of the extracts (no difference between conventional and biological treatment), indicating that the conventional treatment did not still affect the stable nucleus of humic fraction.

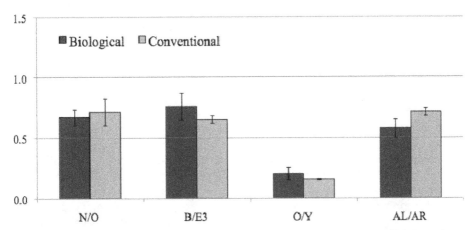

Fig. 7. Indices of mineralization (furfural/pyrrole N/O, pyrrole/phenol O/Y), humification (benzene/toluene B/E_3), and aliphatic/aromatic (AL/AR) ratios (AL=E_1+K+N, AR=B+O+E_3+Y) (mean of three replicates ± standard deviation) in bulk soils. The values of O/Y have been divided by 10.

bulk soil	E_1%	K%	B%	O%	E_3%	N%	Y%
Biological	14,3	11,4	16,8	16,4	22,1	11,0	8,06
Conventional	21,0	11,0	14,3	13,6	21,9	9,70	8,63

Table 11. Relative abundances (%) of main pyrolytic peaks (acetonitrile E_1, acetic acid K, benzene B, pyrrole O, toluene E_3, furfural N, phenol Y) in bulk soils.

Soil extract	E1%	K%	B%	O%	E3%	N%	Y%
Biological	3,38	13,4	21,8	15,6	24,0	13,8	8,02
Conventional	6,00	15,9	22,8	15,8	23,9	11,5	4,10

Table 12. Relative abundances (%) of main pyrolytic peaks (acetonitrile E_1, acetic acid K, benzene B, pyrrole O, toluene E_3, furfural N, phenol Y) in soil extracts.

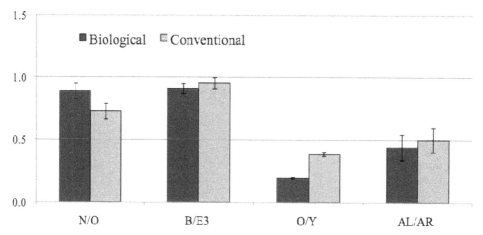

Fig. 8. Indices of mineralization (furfural/pyrrole N/O, pyrrole/phenol O/Y), humification (benzene/toluene B/E_3), and aliphatic/aromatic (AL/AR) ratios (AL=E_1+K+N, AR=B+O+E_3+Y) (mean of three replicates ± standard deviation) in soil extracts. The values of O/Y have been divided by 10.

3.3 Restoration of degraded soils

An experimental field (sandy clay loam soil, USDA classification) in province of Murcia (Spain) was split into plots in which 20 years ago was added a single dose of fresh easily degradable municipal organic waste (MOW) in such dose as to increase the soil organic matter by 0.5 (65t/ha), 1.0 (130 t/ha), 1.5 (200t/ha) and 2.0% (260 t/ha), MOW fraction was incorporated into the top 15 cm of soil using a rotovator. The aim was to restore biochemical and microbial properties and to contrast the erosion through a stimulation of spontaneous establishment of grass cover. One plot without the addition of MOW was used as control (0%). The plots were monitored for three years: following organic amendments an averagely 50-70% plant coverage developed and persisted throughout the experiment until today. The highest plant density were found in 1.5% treatment (80-90%) while only 20-30% plant coverage was developed in the control soil. After 16 years from the treatment the same vegetated plots have been sampled at 0-15 cm depth in order to assess and investigate the changes occurred in consequence of the vegetal cover establishment.

The input of exogenous OM produced qualitative changes of soil OM both in the bulk soil and in the pyrophosphate extracts, which are still evident after 16 years. Considering the different plots, in fact, there were significant differences in the Py-GC fragments analysis: acetonitrile (E_1) and acetic acid (K) increased with the increase of organic matter added (Table 13), at the same time lower values of pyrrole (O) and phenol (Y) were detected in the same plots. The increase of acetonitrile and pyrrole were probably due to the higher cover of vegetation restored in the plots, which bring to higher quantity of cellulose, lipids and easily degradable compounds released into the soil. This trend was also confirmed by AL/AR index, which increased at the increase of the organic matter application (Figure 9).

The plots treated with the higher quantity of organic matter, also showed higher level of humification index (B/E_3), thus suggesting that after 16 years the humification overrides the mineralization process. As suggested by to the humification index, the N/O ratio showed higher values in the plots with the greater application of organic residues, indicating the presence of more evolved (less mineralizable) organic compounds in the soil. The similarity indices confirmed the role of the MOW in affecting the quality of organic matter (Marinari et al., 2007) (Table 14), in that 1%, 1.5% and 2% treatments resulted more similar between them.

bulk soil	E_1%	K%	B%	O%	E_3%	N%	Y%
Organic matter 0.0%	6,39	3,54	11,8	15,1	24,0	14,0	25,2
Organic matter 0.5%	8,81	6,57	12,5	16,9	20,3	16,0	19,5
Organic matter 1.0%	14,4	11,3	15,3	10,6	20,0	10,3	17,4
Organic matter 1.5%	11,9	13,2	14,1	12,7	20,1	14,1	13,9
Organic matter 2.0%	14,6	11,6	14,7	10,7	19,9	12,7	15,7

Table 13. Relative abundances (%) of main pyrolytic peaks (acetonitrile E_1, acetic acid K, benzene B, pyrrole O, toluene E_3, furfural N, phenol Y) in bulk soils.

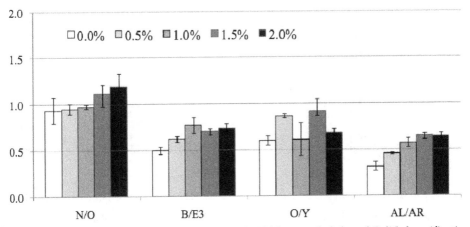

Fig. 9. Indices of mineralization (furfural/pyrrole N/O, pyrrole/phenol O/Y), humification (benzene/toluene, B/E_3), and aliphatic/aromatic (AL/AR) ratios (AL=E_1+K+N, AR=B+O+E_3+Y) (mean of three replicates ± standard deviation) in bulk soils.

bulk soil	0,5%	1%	1,5%	2%
0%	0,800	0,641	0,695	0,660
0,5%		0,736	0,779	0,747
1.0%			0,853	0,945
1,5%				0,898

Table 14. Pyrolitic index of similarity in the bulk soils.

The results of the humic extract highlighted the great difference among the 1.5% and the others treatments (Table 15,16 and Figure 10). In fact, the higher vegetal cover in 1.5% plots probably promoted the enrichment of soil in labile and less stable fraction of organic matter, which resulted the more sensitive to degradation process, as suggested by the trend of N/O and O/Y. As a consequence, in 1.5% treatment a lower value of B/E_3 was observed, probably due to the masking of humic matter from fresh materials.

Soil extract	$E_1\%$	K%	B%	O%	$E_3\%$	N%	Y%
Organic matter 0.0%	3,66	3,94	16,4	13,2	21,2	18,7	22,9
Organic matter 0.5%	2,62	2,40	17,9	12,5	21,5	19,2	24,0
Organic matter 1.0%	3,46	3,77	18,1	12,0	22,5	17,2	23,2
Organic matter 1.5%	7,88	6,05	12,7	14,9	22,2	17,1	19,0
Organic matter 2.0%	6,07	4,93	15,6	12,2	23,6	15,9	21,7

Table 15. Relative abundances (%) of main pyrolytic peaks (acetonitrile E_1, acetic acid K, benzene B, pyrrole O, toluene E_3, furfural N, phenol Y) in soil extracts.

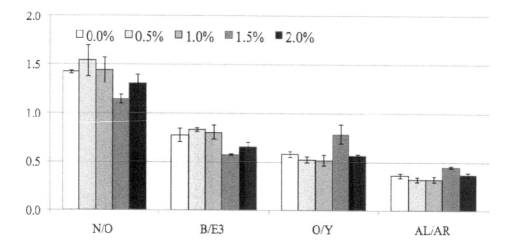

Fig. 10. Indices of mineralization (furfural/pyrrole N/O, pyrrole/phenol O/Y), humification (benzene/toluene B/E_3), and aliphatic/aromatic (AL/AR) ratios (AL=E_1+K+N, AR=B+O+E_3+Y) (mean of three replicates ± standard deviation) in soil extracts. The values of O/Y have been divided by 10.

Soil extract	0,5%	1%	1,5%	2%
0%	0,872	0,938	0,781	0,888
0,5%		0,88	0,703	0,773
1.0%			0,766	0,855
1,5%				0,852

Table 16. Pyrolitic index of similarity in the soil extracts.

4. Statistical considerations

In order to understand the importance of PY-GC technique in describing the chemico-structural and the functionality of soil organic matter, the principal component analysis (PCA) was used. PCA is a multivariate statistical data analysis technique that reduces a set of raw data into a number of principal components that retain the most variance within the original data to identify possible patterns or clusters between objects and variables.

In order to develop wider conclusions about the state of soil organic matter a data set containing information about different ecosystems was used. 12 sites were selected across Europe in order to ensure a broad range of different conditions (natural soil and agro-ecosystems) and related pressures: Spain (Abanilla, Santomera, El Aguilucho, Tres Caminos, all in Murcia region) Italy, (Castelfalfi and Alberese in Tuscany region, Pantanello in Basilicata region), Germany (Puch, in Bavaria region) and Hungary (Gödöllö). All selected sites were characterized by local variation of anthropogenic and natural pressures additionally to the variation that can be found at the European scale. The details of all sites are listed in Table 17. The selected sites may be grouped onto two different types:

Type one: Catenae. Pressures and soil degradation change along a gradient, for example along a slope with differing steepness or along a landscape with a sequence of plant cover. The selected catenae imply a variation of pressures that is mainly of natural origin. Samples are taken from different locations along the catenae. The sites were situated in Spain (Abanilla, Carcavo, Santomera) and Hungary (Godollo) and are described in Table 17.

Site	Description	Treatment Pression	Experiment
Puch (Germany)	Agricultural field under normal cultivation	Tillage/ Plant Cover	Management variation
	Field kept bare of vegetation with plowing		
	Very limited vegetation maintained		
Abanilla (Spain)	Bare	Erosion	Catena
	Esparto grass		
	High coverage		
	Medium coverage		
	Sludge added 0%	Organic matter	Management variation
	Sludge added 5%		
	Sludge added 10%		
	Sludge added 15%		
	Sludge added 20%		

Site	Description	Treatment Pression	Experiment
Carcavo (Spain)	Abandoned	Erosion/ Exposition	Catena
	Degraded		
	Vegetated with pines		
	North slope		
	South slope		
El Aguilucho (Spain)	Control	Plant Cover/ Erosion/ Organic Matter	Management variation
	Pine		
	Pine + Organic Matter (OM)		
	Terrace + Pine		
	TP+ mycorrize		
	TPm + OM		
	TP+ soil mycorrize		
	TP + OM		
	TPMs + OM		
Santomera (Spain)	Bare	Plant Cover/ Erosion	Management variation/ Catena
	Forest		
	Shrub		
	Removed forest		
	Forest		
Tres Caminos (Spain)	Compost	Plant Cover/ Organic matter	Management variation
	Humus enzymes		
	Control		
	Reforestation		
	Reforestation + mycorrize		
	Seed		
	Sewage Sludge		
Gödöllö (Hungary)	Accumulated	Erosion	Catena
	High erosion		
	Medium erosion		
	No erosion		
Pantanello (Italy)	Organic agriculture	Organic matter/ Agrochemicals	Management variation
	Conventional agriculture		
Castelfalfi (Italy)	Disturbed forest	Animal pressure/ Organic matter/ Erosion	Management variation
	Undisturbed forest		
Alberese (Italy)	Organic agriculture	Organic matter/ Agrochemicals	Management variation
	Conventional agriculture		

Table 17. Site description

Type two: Variation of management. Different types of treatments have been applied to plots at the same site some time ago. These treatments may either represent an anthropogenic pressure, which promotes soil degradation or a remediation action, from which lasting effects may be expected. Type two also contains adjacent sites under different agricultural and forest management, where for instance organic farming may be compared to common farming practice or fenced off forest areas may be compared to the rest of the forest with a high animal population density. The sites were situated in Germany (Puch), Spain (Abanilla, El Aguilucho, Santomera, Tres Caminos) and Italy (Pantanello, Castelfalfi and Alberese) and are described in Table 17.

The PCA of the data set about bulk soils indicated 80% of the data variance as being contained in the first three components (Table 18). PC1, PC2 and PC3 account for 31.0%, 27.9% and 22.0% of the total variance, respectively. PC1 was closely associated with B, O, Y and both indices of mineralization (O/Y and N/O), while PC2 was linked with K, E_3, humification index B/E_3 and AL/AR index. PC3 was associated with E_1 and N. As expected, the indices of mineralization had an opposite behaviour: when O/Y increase, N/O decrease and vice versa, being involved in pseudo-stable and labile organic matter metabolism, respectively. In PC2, the high correlation occurred between humification index B/E_3 and AL/AR index, suggested that both labile and stable organic matter affected soil humification processes.

	Factor 1	Factor 2	Factor 3
Acetonitrile (E_1 soil)	0.027	0.139	0.899*
Acetic acid (K soil)	0.072	0.875*	0.190
Benzene (B soil)	0.730*	-0.014	0.285
Pyrrole (O soil)	0.804*	0.020	-0.483
Toluene (E_3soil)	0.071	-0.819*	0.367
Furfural (N soil)	-0.257	0.136	-0.872*
Phenol (Ysoil)	-0.786*	-0.283	-0.352
Mineralization index (N/O soil)	-0.788*	0.031	-0.321
Humification index (B/E_3 soil)	0.290	0.868*	-0.188
Mineralization index (O/Y soil)	0.895*	0.012	-0.165
Energetic reservoir index (AL/AR soil)	-0.137	0.871*	0.375
Explained .Variance	3.398	3.068	2.482
Total Proportionality	30.89%	27.89%	22.57%

Table 18. Analysis of principal components on bulk soil pyrolytic fragments and indices. *Variables with component loading used to read the PC.

The results of PCA of the data set about soil extracts were similar to those about bulk soil, with few important exceptions concerning pyrolytic fragments and index related to the more stable part of humic matter. In the same way of bulk soil, PCA of the extract isolated three principal components (PC) (total variance explained: 70.6%). The 1st PC (29.9 % of the total variance) included E_3 and the same parameters of bulk soil (O, Y, O/Y and N/O) with the exception of B that, in this case, was included in the 3nd PC (17.1 % of the total variance) together with humification index (B/E_3). Therefore, differently from the bulk soil, the humification index resulted independent from the other indices, suggesting the importance of the PY-GC on the pyrophosphate extract to investigate the condensed aromatic nucleus of humic substances. The 2nd PC(23.7% of the total variance) was associated, as for bulk soil, with K and AL/AR index (Table 19).

	Factor 1	Factor 2	Factor 3
Acetonitrile (E$_1$ extract)	0.074	-0.479	0.203
Acetic acid (K extract)	-0.089	-0.886*	-0.225
Benzene (B extract)	-0.256	0.358	-0.804*
Pyrrole (O extract)	-0.742*	0.260	0.109
Toluene (E$_3$ extract)	-0.749*	0.352	0.207
Furfural (N extract)	0.271	0.319	0.494
Phenol (Y extract)	0.771*	0.485	0.134
Mineralization index (N/O extract)	0.742*	0.056	0.346
Humification index (B/E$_3$ extract)	0.355	-0.001	-0.833*
Mineralization index (O/Y extract)	-0.859*	-0.102	-0.074
Energetic reservoir index (AL/AR extract)	0.083	-0.960*	0.067
Explained. Variance	3.279	2.604	1.879
Total Proportionality	29.80%	23.67%	17.08%

Table 19. Analysis of principal components on humic extract pyrolytic fragments and indices. *Variables with component loading used to read the PC.

5. Conclusions

The results obtained by PY-GC carried out on the bulk soil and humic extract of several European sites, characterized by different management practices, permitted to draw these final considerations:

i.   Mineralization indices N/O and O/Y resulted the pyrolytic parameters that better discriminate among the different soils, being involved in the dynamics of labile organic matter, linked to microbial metabolism, and pseudo-stable matter, related to the humic fraction.

ii.   The pyrolysis on the humic extract often evidenced information hidden by the analysis carried out on the bulk soil. In particular, the pyrolysis on the pyrophosphate extract permitted to obtain clearer information on the condensed aromatic nucleus of humic substances.

iii. Even if the analysis on soil as such (bulk) and on soil humic extracts gave each specific information on the quality of the organic matter; the scientific details obtained from the pyrophosphate extract agreed with those resulted analysing the bulk soil, meaning that this kind of extract does not alter the chemical properties of the soil organic matter.

Finally, it is possible to conclude that the PY-GC can be used as a quick and general reproducible technique to make a qualitative study of the chemico-structural characteristics of soil organic matter turnover related to biological activity (bulk soil), and of the more stable humic fraction (pyrophosphate extract) in soils under different anthropogenic and natural pressures.

Further researches could involve the study of the link between the chemico-structural properties of the humic fraction and their capability to protect enzyme in biologically active form.

6. Acknowledgment

The study was carried out within the framework of the EU project "Indicators and thresholds for desertification, soil quality, and remediation" INDEX (STREP contract n° 505450 2004/2006).

We would like to thank Maurizio Calcinai for his assistance in the pyrolysis-gas chromatography analysis.

7. References

Alberts, J.J.& Takacs, M. (2004). Total luminescence spectra of IHSS standard and reference fulvic acids, humic acids and natural organic matter: comparison of aquatic and terrestrial source terms. *Organic Geochemistry*, Vol. 35, pp. 243–256

Alcaniz, J.M., Seres, A. & Gassiot-Matas, M. (1984). Diferenciacion entre humus mull carbonatado y mull evolucionado por Py-GC. In: *Proceedings of the National Conference Ciencia Del Suelo*, 217-228, Madrid.

Bandick, A.K. & Dick, R.P. (1999). Field management effects on soil enzyme activities. *Soil Biology & Biochemistry*, Vol. 31, pp. 1471-1479

Bracewell, J.M. & Robertson, G.W. (1984). Quantitative comparison of nitrogen containing pyrolytic products and aminoacids composition of soil humic acids. *Journal of Analytical and Applied Pyrolysis*, Vol. 6, pp. 19–29

Buurman, P., Peterse, F. & Almendros Martin, G. (2007). Soil organic matter chemistry in allophanic soils: A pyrolysis-GC/MS study of a costa rican andosol catena. *European Journal of Soil Science*, Vol. 58, pp. 1330-1347

Campo, J., Nierop, K.G.J., Cammeraat, E., Andreu, V. & Rubio, J.L. (2011). Application of pyrolysis-gas chromatography/mass spectrometry to study changes in the organic matter of macro- and microaggregates of a Mediterranean soil upon heating. *Journal of Chromatography A*, Vol. 1218, pp. 4817-4482

Ceccanti, B., Nannipieri, P., Cervelli, S. & Sequi, P. (1978). Fractionation of humus-urease complexes. *Soil Biology & Biochemistry*, Vol. 10, pp. 39–45

Ceccanti, B., Alcaniz, J.M., Gispert, M. & Gassiot, M. (1986). Characterization of organic matter from two different soils by pyrolysis–gas chromatography and isoelectrofocusing. *Soil Science*, Vol. 142, pp. 83–90

Ceccanti, B. & Garcia, C. (1994). Coupled chemical and biochemical methodologies to characterize a composting process and the humic substances. In: Senesi, N., Miano, T.M. (Eds.), *Humic Substances in the Global Environment and Implication on Human Health*, 1279–1284, London

Ceccanti, B. & Masciandaro, G. (2003). Stable humus-enzyme nucleus: the last barrier against soil desertification. In: Lobo, M.C., Ibanez, J.J. (Eds.), *Preserving Soil Quality and Soil Biodiversity – The Role of Surrogate Indicators*, 77–82, CSIC-IMIA, Madrid

Ceccanti, B., Masciandaro, G. & Macci C. (2007). Pyrolysis-gas chromatography to evaluate the organic matter quality of a mulched soil. *Soil & Tillage Research*, Vol. 97, pp. 71-78

Clapp, C.E. & Hayes, M.H.B. (1999). Characterization of humic substances isolated from clay- and silt-sized fractions of a corn residue-amended agricultural soil. *Soil Science*, Vol. 164, pp. 899–913

Dignac, M.F., Houot, S., Francou, C. & Derenne, S. (2005). Pyrolytic study of compost and waste organic matter. Organic *Geochemistry*, Vol. 36, pp. 1054-1071

Doran, J.W. & Parkin, T.B. (1994). Defining and assessing soil quality. In: Doran, J.W., Coleman, D.C., Bezdicek, D.F., Stewart, B.A. (Eds.), *Defining Soil Quality for a Sustainable Environment*. Special Publication, Vol. 35, *Soil Science Society of America*, Inc., 3-21, Madison, WI

European Commission (2006). Soil Thematic Strategy. In: *European Commission Environment Soil*. Last access September 2011. Available from:
http://ec.europa.eu/environment/soil/index_en.htm

Faure, P., Vilmin, F., Michels, R., Jardé, E., Mansuy, L., Elie, M. & Landais, P. (2002). Application of thermodesorption and pyrolysis-GC–AED to the analysis of river sediments and sewage sludges for environmental purpose. *Journal of Analytical and Applied Pyrolysis*, Vol. 62, pp. 297–318

Garcia, C., Hernandez, T. & Costa, F. (1992). Variation in some chemical parameters and organic matter in soils regenerated by the addition of municipal solid waste. *Environmental Management*, Vol. 16, pp. 763-768

Garcia, C., Hernandez, T., Costa, F., Ceccanti, B. & Calcinai, M. (1992). A chemical-structural study of organic wastes and their humic acids during composting by means of pyrolysis-gas chromatography. *Science of the Total Environment*, Vol. 119, pp. 157-168

Gondar, D., Lopez, R., Fiol, S., Antelo, J.M. & Arce, F. (2005). Characterization and acid–base properties of fulvic and humic acids isolated from two horizons of an ombrotrophic peat bog. *Geoderma*, Vol. 126, pp. 367–374

Grandy, A.S. & Neff, J.C. (2008). Molecular C dynamics downstream: The biochemical decomposition sequence and its impact o soil organic matter structure and function. *Science of the Total Environment*, Vol. 404, pp. 297-307

Lobe, I., Du Preez, C.C. & Amelung, W. (2002). Influence of prolonged arable cropping on lignin compounds in sandy soils of the south african highveld. *European Journal of Soil Science*, Vol. 53, pp. 553-562

Marinari, S., Liburdi, K., Masciandaro, G., Ceccanti, B. & Grego, S. (2007a). Humification-mineralization pyrolytic indices and carbon fractions of soil under organic and conventional management in central Italy. *Soil & Tillage Research*, Vol. 92, pp.10-17

Marinari, S., Masciandaro, G., Ceccanti, B. & Grego, S. (2007b). Evolution of soil organic matter changes using pyrolysis and metabolic indices: A comparison between organic and mineral fertilization *Bioresource Technology*, Vol. 98, pp. 2495-2502

Masciandaro, G., Ceccanti, B. & Garcia, C. (1997). Soil agroecological management: fertirrigation and vermicompost treatments. *Bioresource Technology*, Vol. 59, pp. 199-206

Masciandaro, G. & Ceccanti, B. (1999). Assessing soil quality in different agroecosystems through biochemical and chemico-structural properties of humic substances. *Soil & Tillage Research*, Vol. 51, pp. 129-137

Nannipieri, P., Muccini, L. & Ciardi, C. (1983). Microbial biomass and enzyme activities: production and persistence. *Soil Biology & Biochemistry*, Vol. 15, pp. 679-685

Nierop, K.G.J. (2001). Temporal and vertical organic matter differentiation along a vegetation succession as revealed by pyrolysis and thermally assisted hydrolysis and methylation. *Journal of Analytical and Applied Pyrolysis*, Vol. 61, pp. 111-132

Nierop, K.G.J., Pulleman, M.M. & Marinissen, J.C.Y. (2001). Management-induced organic matter differentiation in grassland arable soil: a study using pyrolysis techniques. *Soil Biology & Biochemistry*, Vol. 33, pp. 755-764

Peruzzi, E., Masciandaro, G., Macci, C., Doni, S., Mora Ravelo, S.G., Peruzzi, P. & Ceccanti, B. (2011). Heavy metal fractionation and organic matter stabilization in sewage sludge treatment wetlands. *Ecological Engineering*, Vol. 37, pp. 771-778

Piccolo, A. (1996). Humus and soil conservation. In: Humic Substances in Terrestrial Ecosystems (ed. A. Piccolo), 225-264, Elsevier, Amsterdam

Reeves, D.W. (1997). The role of soil organic matter in maintaining soil quality in continuous cropping systems. *Soil & Tillage Research*, Vol. 43, pp. 131-167

Schnitzer, M. (1991). Soil organic matter. The next 75 years. *Soil Science*, 151, pp. 41-58

See, J.H. & Bronk, D.A. (2005). Changes in C:N ratios and chemical structures of estuarine humic substances during aging. *Marine Chemistry*, Vol. 97, pp. 334-346

Sobeih, K.L., Baron, M. & Gonzalez-Rodriguez, J. (2008). Recent trends and developments in pyrolysis-gas chromatography. *Journal of Chromatography A*, Vol. 1186, pp. 51-66

Sollins, P., Homann, P. & Caldwell, B.A. (1996). Stabilization and destabilization of soil organic matter: Mechanisms and controls. *Geoderma*, Vol. 74, pp. 65-105

Van Bergen, P.F., Nott, C.J., Bull, I.D., Poulton, P.R. & Evershed, R.P. (1998). Organic geochemical studies of soils from the rothamsted classical experiments - IV. preliminary results from a study of the effect of soil pH on organic matter decay. *Organic Geochemistry*, Vol. 29, pp. 1779-1795

Vancampenhout, K., De Vos, B., Wouters K., Van Calster H., Swennen R., Buurman P. & Deckers J. (2010). Determinants of soil organic matter chemistry in maritime temperate forest ecosystems. *Soil Biology & Biochemistry,* Vol. 42, pp. 220-233

GC Analysis of Volatiles and Other Products from Biomass Torrefaction Process

Jaya Shankar Tumuluru[1,*], Shahab Sokhansanj[2],
Christopher T. Wright[1] and Timothy Kremer[3]
*[1]Biofuels and Renewable Energies Technologies,
Idaho National Laboratory, Idaho Falls,*
*[2]Bioenergy Resource and Engineering Systems Group, Environmental Sciences Division,
Oak Ridge National Laboratory, Oak Ridge,*
*[3]Chemical and Biomolecular Engineering Department,
The Ohio State University, Columbus,
USA*

1. Introduction

Gas chromatography (GC) is a common method used to analyze gases produced during various chemical processes. Torrefaction, for example, is a method for pretreating biomass to make it more suitable for bioenergy applications that uses GC to characterize products formed during the process. During torrefaction, biomass is heated in an inert environment to temperatures ranging between 200–300°C. Torrefaction causes biomass to lose low-energy condensables (liquids) and non-condensable volatiles, initially in gas form, thereby making biomass more energy dense.

The condensable volatiles are divided into three subgroups: reaction water, organics, and lipids. The first subgroup is reaction water, which contains free and bound water released from the biomass by evaporation. The second subgroup consists of organics (in liquid form) that are mainly produced during devolatilization, depolymerization and carbonization reactions in the biomass. The final subgroup, lipids, consists of compounds that are present in the original biomass, such as waxes, fatty acids, and non-condensable gases — mainly CO_2, CO, and small amounts of methane. Knowing the composition of volatiles produced in the torrefaction temperature range can shed light on the raw material adaption process, process control, process behavior and operation, energy process and energy optimization, and green chemical production.

In general, GC with mass spectroscopy is used for both condensable and non-condensable gases. GC configuration plays an important role in accurately identifying the compounds in these gases. A combination of different detectors — like thermal, flame, and photo ionization detectors — combined with mass spectrometers are used for profiling both condensable and non-condensable gases.

---

* Corresponding Author

2. Gas Chromatography

Chromatography is an important analytical tool that allows for the separation of components in a gas mixture. GC is a common type of chromatography used to separate and analyze compounds that can be vaporized without decomposition. Typical uses of GC include testing the purity of a particular substance or separating the different components and relative amounts of different components of a mixture. GC can also be used to prepare pure compounds from a mixture (Pavia et al., 2006). In GC, the mobile phase (or "moving phase") is a carrier gas, usually an inert gas such as helium or an unreactive gas such as nitrogen. The stationary phase is a microscopic layer of liquid or polymer on an inert solid substrate inside a glass or metal tube called a column. The gases are analyzed as they interact with the column walls, which have been coated with different stationary phases. This coating results in the compounds eluting at different times, called the retention time for each compound. These compounds are then further analyzed by comparison with calibrated standard gases (Pavia et al., 2006).

The working principle of GC is similar to that of fractional distillation, as both processes use temperature to separate the components. The only difference is that distillation is used in large-scale applications, where GC is only used in smaller-scale applications (Pavia et al., 2006). Because of its mechanisms for analysis, GC is also sometimes known as vapor-phase chromatography (VPC), or gas–liquid partition chromatography (GLPC) (Pavia et al., 2006).

3. GC components

Figure 1 shows the typical components of GC system: the sample injector, column, detector, and carrier gases.

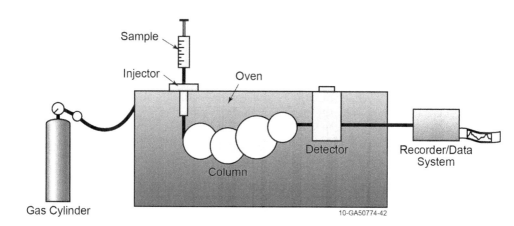

Fig. 1. Schematic view of gas chromatography (Punrattanasin and Spada, 2011)

3.1 Types of samplers

Manual insertion of a sample in a GC system for analysis is possible, but is no longer preferred for reasons of accuracy. In fact, manual injection is avoided where there is a need for high accuracy and precise results. A better option is an auto-sampler that injects samples automatically into the inlets of the GC system. Automatic injection helps achieve consistency and reproducibility and reduces analysis time (Pavia et al., 2006). The several types of automatic samplers are:

- Liquid injection
- Static head-space by syringe technology
- Dynamic head-space by transfer-line technology
- Solid phase microextraction (SPME).

3.2 Types of inlets

An inlet is hardware attached to the column head that introduces the samples into the continuous flow carrier gas. Common types of inlets are split/splitless (S/SL), on-column, programmed temperature vaporization (PTV) injectors, gas switching valve, and purge and trap systems.

An S/SL injector introduces a sample into a heated chamber via a syringe through a septum, and the heat helps volatilization of the sample and sample matrix. In split mode, the sample/carrier gas is exhausted into the chamber through the split vent, and this is preferred when the sample has high analyte concentrations (>0.1%). In the case of the splitless mode, the valve opens after a pre-set time to purge heavier elements in order to prevent the contamination of the system. An on-column inlet is a simple component through which the sample is introduced directly into the column without heat. A PTV injector introduces a sample into the liner at a controlled injection rate. The temperature of the liner is usually slightly below the boiling point of the solvent so that the solvent is continuously evaporated and vented through the split line. PTV injectors are particularly suited to in-liner derivatisation due to the flexibility of control over parameters such as injection volume (Bosboom et al., 1996), carrier gas flow and liner temperature (Poy et al., 1981). A gas-source inlet (or gas-switching valve) injects gaseous samples into the collection bottles through a six-port switching valve. Upon switching, the contents of the sample loop are inserted into the carrier gas stream. Finally, the purge-and-trap system purges insoluble volatile chemicals from the matrix by bubbling inert gas through an aqueous sample. The volatiles are 'trapped' on an adsorbent column (also known as a trap or concentrator) at ambient temperature. The trap is then heated, and the volatiles are directed into the carrier gas stream (Pavia et al., 2006; Kaufmann, 1997).

3.3 Types of detectors

A number of detectors may be selected in gas chromatography, based on the sample to be separated. The most common are the flame ionization detector (FID) and the thermal conductivity detector (TCD). These detectors are sensitive to a wide range of components and work for wide-ranging concentrations. TCDs are essentially universal and can be used to detect any component other than the carrier gas so long as the thermal conductivities are different from those of the carrier gas at a specific detector temperature. TCD has a detection limit of ~100 ppm (Goorts, 2008).

FIDs are more sensitive than TCDs and have a detection limit of ~1 ppm (Goorts, 2008), but their major limitation is the number of molecules they can successfully detect. FIDs can only detect combustible carbon atoms; therefore, FIDs are primarily used for detecting organic molecules. FIDs cannot detect H_2O or CO_2, so they are undesirable for some applications (Goorts, 2008). As TCD analysis is non-destructive, it can be operated in-series before an FID, which is destructive, thus providing complementary detection of the same analyte (Goorts, 2008).

Other detectors used for specific types of substances or narrower concentration ranges are

- Catalytic combustion detector (CCD), which measures combustible hydrocarbons and hydrogen
- Discharge ionization detector (DID), which uses a high-voltage electric discharge to produce ions
- Dry electrolytic conductivity detector (DELCD), which uses an air phase and high temperature (v. Coulsen) to measure chlorinated compounds
- Electron capture detector (ECD), which uses a radioactive Beta particle (electron) source to measure the degree of electron capture
- Flame photometric detector (FPD)
- Hall electrolytic conductivity detector (ElCD)
- Helium ionization detector (HID)
- Nitrogen phosphorus detector (NPD)
- Infrared detector (IRD)
- Mass selective detector (MSD)
- Photo-ionization detector (PID)
- Pulsed discharge ionization detector (PDD)
- Thermal energy (conductivity) analyzer/detector (TEA/TCD)
- Thermionic ionization detector (TID).

The various detectors, support gases, selectivity, detectability and dynamic range are given in Table 1 (SHU, 2011).

Some gas chromatographs are connected to a mass spectrometer (MS), nuclear magnetic resonance (NMR) spectrometer, or infrared (IR) spectrophotometer that acts as the detector. These combinations are known as GC-MS, GC-MS-NMR, and GC-MS-NMR-IR. However, it is very rare to use the NMR and IR along with GC. The most commonly used is GC-MS (Scott, 2003).

4. Carrier gas

The carrier gas plays an important role in GC analysis and can vary depending on the GC used. Carrier gas must be dry, free of oxygen, and chemically inert. Typical carrier gases include helium, nitrogen, argon, hydrogen, and air. The choice of carrier gas (mobile phase) is important, with hydrogen providing the best separation. Helium is most commonly used as a carrier gas because it has larger flow rates, is non-flammable, and can work with many detectors. The type of carrier gas used depends on the following factors: (a) detector used; (b) sample's composition; (c) safety and availability (i.e., hydrogen is flammable, high-purity helium can be difficult to obtain in some areas of the world); and (d) purity of the carrier gas (where high pure gases of 99.995% are selected when high sensitivities in the measurement are required) (Guiochon and Guillemin, 1988).

Detector	Type	Support Gases	Selectivity	Detectability	Dynamic Range
Flame ionization (FID)	Mass flow	Hydrogen and air	Most organic compounds	100 pg	10^7
Thermal conductivity (TCD)	Concentration	Reference	Universal	1 ng	10^7
Electron capture (ECD)	Concentration	Make-up	Halides, nitrates, nitriles, peroxides, anhydrides, and organic metals	50 fg	10^5
Nitrogen phosphorous	Mass flow	Hydrogen and air	Nitrogen, phosphorous	10 pg	10^6
Flame photometric (FPD)	Mass flow	Hydrogen and air, possibly oxygen	Sulphur, phosphorous, tin, boron, arsenic, germanium, selenium, and chromium	100 pg	10^3
Photo-ionization (PID)	Concentration	Make-up	Alaphatics, aromatics, ketones, esters, aldehydes, amines, hetrocyclics, organosulphurs, some organometallics	2 pg	10^7
Hall electrolytic conductivity	Mass flow	Hydrogen, oxygen	Halides, nitrogen, nitroamine, sulphur		
Note: pg: picogram; ng: nanogram, and fg: femtogram					

Table 1. Various detectors and their detectability

4.1 GC methods for sample analysis

An analysis method comprises conditions in which a GC operates for a specific analysis. Developing the method involves finding the conditions required for the analysis of a specific sample. The process conditions that can be used to generate the method file are (a) inlet temperature, (b) detector temperature, (c) column temperature, (d) temperature program, (e) carrier gases and their flow rates, (f) the column's stationary phase, (g) column diameter and length, (h) inlet type and flow rates, and (i) sample size and injection technique. These operating conditions depend on the detector selected and the compounds to be analyzed (Grob et al., 2004).

4.2 Gas Chromatography-Mass Spectrometry (GC-MS)

GC-MS uses two techniques that are combined into a single method for analyzing mixtures of chemicals. Gas chromatography separates the components of a mixture, and mass spectroscopy characterizes each of the components individually. Combining the two techniques helps to analyze the samples both qualitatively and quantitatively. As the sample is injected into the chromatograph, the sample mixture gets separated into individual components due to different flow rates. This results in quantitative analysis of the components, along with a mass spectrum of each component. Applications of GC-MS include drug detection, fire investigation, environmental analysis, explosives investigation, and identification of unknown samples. Strengths of GC/MS analysis are (a) identification of organic components from complex mixtures, (b) quantitative analysis, and (c) determination of traces of organic contamination. A major limitation of GC/MS is that the sample must be volatile or capable of derivatization (Hübschmann, 2008).

5. Volatile analysis during biomass torrefaction

5.1 Biomass as an energy source

Utilization of biomass for energy can help reduce world dependence on fossil fuels, reduce the impact of greenhouse gasses on the environment, and help meet targets established in the Kyoto Protocol (UN, 1998). Biomass usage as a renewable energy resource is increasing as it is considered carbon neutral (carbon dioxide released is already part of the carbon cycle). Commercial limitations of biomass are its low bulk density, high moisture content, hydrophilic nature, and low calorific value (Arias et al., 2008). Due to its low energy content compared to fossil fuels, it may be necessary to use high volumes of biomass. This results in storage, transportation, and feed-handling problems at bioenergy conversion facilities. These drawbacks have led to the development of new technologies to successfully use biomass for fuel applications. One way to overcome these limitations is pretreating the biomass. There are different pre-treatment methods, including thermal, mechanical, and chemical. Torrefaction is a thermal method that significantly improves the physical and chemical properties of biomass. These compositional changes make torrefaction a promising pre-treatment method for thermochemical conversion and combustion applications (Tumuluru et al., 2011).

5.2 Biomass components

Fig. 2 indicates the low-molecular and macromolecular weight substances in biomass (Mohan et al., 2006). The various components of the biomass are significantly influenced by thermal treatments (Tumuluru et al., 2011).

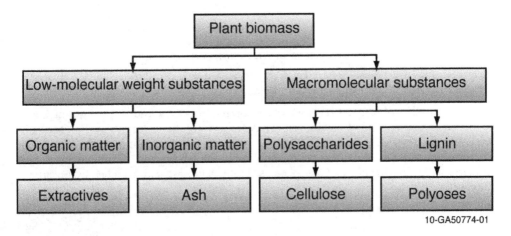

Fig. 2. Plant biomass composition (Mohan et al. 2006)

5.2.1 Cellulose

Cellulose is a high-molecular-weight polymer which provides structural rigidity to the plants. Cellulose degradation begins at 240°C, resulting in anhydrous cellulose and levoglucosan polymer, and makes up the fibers in wood and levoglucosan (Mohan et al., 2006). The crystalline structure resists thermal depolymerization during thermal treatment to a greater degree than is seen in unstructured hemicelluloses.

5.2.2 Hemicellulose

Hemicelluloses are branched polymers which account for about 25–35 wt% in biomass. Thermal degradation of hemicelluloses takes place from 130–260°C, with the majority of weight loss occurring above 180°C (Demibras, 2009; Mohan et al., 2006). Degradation of hemicellulose results in emission of volatiles.

5.2.3 Lignin

Lignin is an amorphous, highly branched, cross-linked macromolecular polyphenolic resin which is covalently linked to hemicellulose and cross-linked with different plant polysaccharides. These linkages give mechanical strength to the cell wall. It is relatively hydrophobic and aromatic in nature. Lignin decomposes when heated above 280°C, producing phenols due to cleavage of ether bonds (Demibras, 2009; Mohan et al., 2006). Lignin converts into char at temperatures >300°C.

5.2.4 Extractables

The other organic extractables present in biomass are fats, waxes, alkaloids, proteins, phenolics, simple sugars, pectins, mucilages, gums, resins, terpenes, starches, glycosides, saponins, and essential oils (Mohan et al., 2006).

6. Torrefaction process

Torrefaction is a process of heating biomass slowly in an inert atmosphere to a maximum temperature of 300°C (Fonseca et al., 1998) and has been defined as the partially controlled and isothermal pyrolysis of biomass occurring in a temperature range of 200–300°C (Bergman and Kiel, 2005). The treatment yields a solid, uniform product with lower moisture and higher energy content compared to raw biomass (Sadaka and Negi, 2009).

The initial heating of biomass during torrefaction removes unbound water. Heating above 160°C removes the bound water due to the thermo-condensation process (Zanzi et al., 2002). Increasing the temperatures to 180–270°C initiates the decomposition of hemicellulose. At temperatures >280°C, the process is completely exothermic and results in significant increase in the production of CO_2, phenols, acetic acid, and other higher hydrocarbons (Zanzi et al., 2002).

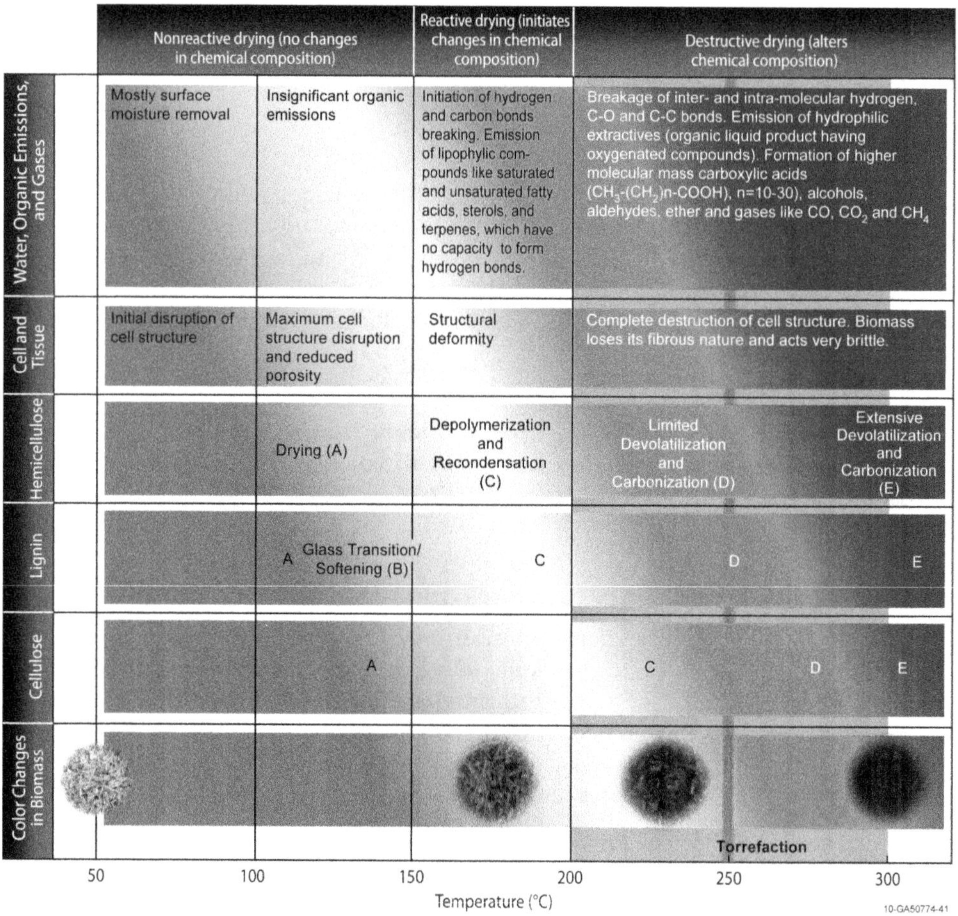

		Nonreactive drying (no changes in chemical composition)		Reactive drying (initiates changes in chemical composition)	Destructive drying (alters chemical composition)	
Water, Organic Emissions, and Gases		Mostly surface moisture removal	Insignificant organic emissions	Initiation of hydrogen and carbon bonds breaking. Emission of lipophylic compounds like saturated and unsaturated fatty acids, sterols, and terpenes, which have no capacity to form hydrogen bonds.	Breakage of inter- and intra-molecular hydrogen, C-O and C-C bonds. Emission of hydrophilic extractives (organic liquid product having oxygenated compounds). Formation of higher molecular mass carboxylic acids $(CH_3\text{-}(CH_2)n\text{-}COOH)$, n=10-30), alcohols, aldehydes, ether and gases like CO, CO_2 and CH_4	
Cell and Tissue		Initial disruption of cell structure	Maximum cell structure disruption and reduced porosity	Structural deformity	Complete destruction of cell structure. Biomass loses its fibrous nature and acts very brittle.	
Hemicellulose			Drying (A)	Depolymerization and Recondensation (C)	Limited Devolatilization and Carbonization (D)	Extensive Devolatilization and Carbonization (E)
Lignin			A Glass Transition/ Softening (B)	C	D	E
Cellulose			A	C	D	E
Color Changes in Biomass					Torrefaction	

50   100   150   200   250   300
Temperature (°C)

10-GA50774-41

Fig. 3. Structural, chemical, and color changes in biomass at different drying temperatures (Tumuluru et al., 2011)

6.1 Biomass reactions

The physiochemical and structural changes in biomass at different temperature regimes is given in Fig. 3 (Tumuluru et al., 2011). The figure indicates that at higher temperature regimes, the drying is more destructive in terms of breakage of inter- and intra-molecular, hydrogen, C-O, and C-C bonds and leads to emission of some hydrophilic, oxygenated compounds. In addition, these changes lead to formation of higher-molecular-mass carboxylic acids, aldehydes, ethers and some gases like carbon monoxide, carbon dioxide and methane.

6.2 Torrefaction gas composition

The three main products produced during torrefaction are a brown/dark-color solid; a condensable liquid, including mostly moisture, acetic acid, and other oxygenates; and non-condensable gases — mainly CO_2, CO, and small amounts of methane. The last two products can be represented by volatiles. Fig. 4 (Bergman et al., 2005) gives an overview of the torrefaction products, classified based on their state at room temperature, which can be,

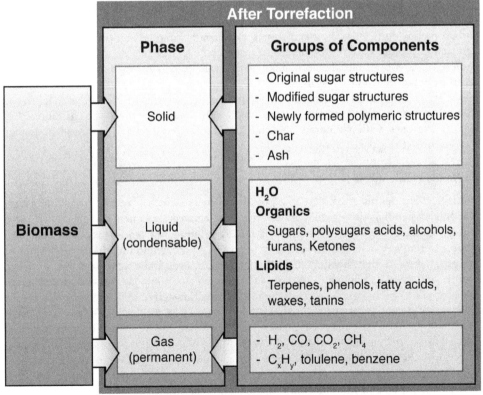

10-GA50774-03

Fig. 4. Products formed during the torrefaction of biomass (Bergman et al. 2005)

solid, liquid, or gas. The solid phase consists of a chaotic structure of the original sugar structures and reaction products. The gas phase includes the gases that are considered permanent gases, but also light aromatic components such as benzene and toluene. The condensables, or liquids, can be divided into three subgroups which include water, organics, and lipids, as shown in Fig. 4. One subgroup is reaction water as a product of thermal decomposition. The liquid also contains the free and bound water that has been released from the biomass by evaporation and dehydration reactions. The organics subgroup (in liquid form) consists of organics that are mainly produced during devolatilization and carbonization. Finally, the lipids are a group of compounds that are present in the original biomass. This subgroup contains compounds such as waxes and fatty acids. The type and amount of the gases that come out depend on the raw material type and torrefaction process conditions, including process temperature and residence time (Tumuluru et al., 2011).

Currently, there is much emphasis on understanding the torrefaction gas composition, as its energy content plays a major role in improving the overall efficiency. Some researchers like Bergman et al. (2005, 2005a), Prins et al. (2006), and Deng et al. (2009) have investigated the torrefaction gas composition.

6.3 Non-condensable gases

Carbon monoxide (CO) is the main source of calorific value among the non-condensable torrefaction products. The formation of CO_2 may be explained by decarboxylation of acid groups in the wood, but the formation of CO cannot be explained by dehydration or decarboxylation reactions. The increased CO formation reported in literature (White and Dietenberger, 2001) is due to reaction of carbon dioxide and steam with porous char. Traces of hydrogen and methane are also detected in the non-condensable product. The ratio of CO to CO_2 increases with temperature because cellulose and lignin decompose at higher temperatures (Prins et al., 2006).

6.4 Condensable gases

In torrefaction, the major condensable product is water, which is released when moisture evaporates as well as during dehydration reactions between organic molecules. Acetic acid is also a condensable torrefaction product, mainly originating from acetoxy- and methoxy-groups present as side chains in xylose units present in the hemicelluloses fraction. Prins et al. (2006) showed that smaller quantities of formic acid, lactic acid, furfural, hydroxy acetone, and traces of phenol are also present in the volatile component liberated during the decomposition of biomass. As a result, more energy is transferred to the volatiles fraction in the form of combustibles such as methanol and acetic acid. Bridgeman et al. (2008) have provided the energy and mass yields, as well as the volatiles lost during the torrefaction of reed canary grass, wheat straw, and willow at different temperatures, as shown in Table 2. Table 3 indicates the physiochemical composition of pine and torrefied pine at temperatures from 240-290°C. With the increase in torrefaction temperature, fixed carbon in the pine increased while volatiles and moisture content decreased. Tables 3 and 4 indicate that, in the torrefaction range (200–300°C), there is significant change in the volatile concentration within the biomass.

	Temperature (K)			
	503	**523**	**543**	**563**
Reed Canary Grass				
Mass yield (daf)	92.6	84.0	72.0	61.5
Energy yield (daf)	93.5	86.6	77.1	69.0
Volatiles (daf)	7.4	16.0	28.0	38.5
Wheat Straw				
Mass yield (daf)	91.0	82.6	71.5	55.1
Energy yield (daf)	93.5	86.2	78.2	65.8
Volatiles (daf)	9.0	17.4	28.5	44.9
Willow				
Mass yield (daf)	95.1	89.6	79.8	72.0
Energy yield (daf)	96.5	92.7	85.8	79.2
Volatiles (daf)	4.9	10.4	20.2	28
Note: daf: dry ash free basis				

Table 2. Energy and mass yields and volatiles lost during torrefaction of reed canary grass, wheat straw, and willow at different temperatures (Bridgeman et al., 2008)

T^a,°C[b]	p^c	240	250	260	270	290
Fixed carbon, %	20.64	23.55	25.59	25.69	29.38	35.39
Elementary analysis C, % O, %	50.98 42.80	51.14 42.70	51.93 42.18	53.78 40.66	53.57 40.67	58.08 36.40
Pentosans, %	9.61	5.93	5.90	3.10	2.54	1.40
Lignin, %	22.84	24.90	28.72	33.44	39.23	53.47
Extractable, %[d]	14.67	8.19	14.09	19.35	16.49	17.98
Moisture, %[e]	10.80	5.66	4.08	3.96	3.76	3.88
Yield, %	–	86.2	81.8	75.7	66.4	48.8

a. in each case, the mean result given was obtained from a minimum of four different experiments
b. roasting time: 30 minutes
c. native pine
d. neutral-solvent extractable (ethanol, benzene, boiling water)
e. powder samples left at the laboratory atmosphere still had a constant humidity

Table 3. Physiochemical analysis of pine and torrefied pine

6.5 Significance of torrefaction gas

The knowledge of the composition of volatiles produced in the temperature range of torrefaction is a topic of interest for understanding of

- Raw material adaption to the process
- Process control
- Process behavior and operation
- Energy process and energy optimization
- Production of green chemicals.

6.5.1 Improving torrefaction process efficiency

Volatiles produced during torrefaction have relatively low calorific values, but the gas is still flammable. If the gases produced during torrefaction are flammable, they may be combusted and used to heat the torrefaction reactor in a recycle loop as suggested by Bergman et al. (2005), as shown in Fig. 5. The torrefaction gas composition must be studied further to find out if the gas is combustible, if autothermal operation of the reactor is possible, or if an external fuel is required, as well as what torrefaction temperature and residence times give flammable gases and the best efficiencies for the process. Using gas chromatography to identify and quantify compounds in these gases can reveal what portion of the gas is flammable, what flame temperature is required to ignite all the flammable components, and theoretically what heating value the gas would have. With this information, the potential use of these gases as a heat source for the reactor can be further evaluated.

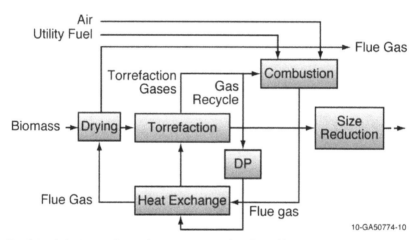

Fig. 5. Combined drying and torrefaction process developed by the Energy Research Centre of the Netherlands (ECN).

6.5.2 Chemical production

Other uses of the condensable fraction of torrefaction gases are for the production of chemicals like concentrated acetic acid, formic acid, methanol, and furfural. The yield of these chemicals depends greatly on torrefaction parameters, including temperature, biomass type, and residence time. In a study of the torrefaction of willow, acetic acid was found to be as much as 5% of the total yield. Formic acid and furfural were as much as 4%, and methanol was at just over 1% (Prins, 2006).

Acetic acid is used mainly to derive more specialized compounds. The largest use of acetic acid is in making vinyl acetate monomer (VAM), which is used in the production of paints, adhesives, paper coatings, and textile finishes. VAM is also used in the production of ethylene vinyl alcohol (EVOH) polymers, which are used in food packing films, plastic bottles, and automobile gasoline tanks. Acetic acid is also used in creating purified terephthalic acid, which is used in making bottle resins and polyester fiber. Another derivative is ethyl acetate, used as a solvent in oil-based lacquers and enamels, including polyurethane finishes and in inks and adhesives. Another large use of acetic acid is in deriving acetate esters, which are used in a wide variety of paints, inks, and other coatings, as well as in many other chemical processes (ICIS News).

Formic acid is a chemical intermediate in the production of various other chemicals and pharmaceuticals, such as caffeine, enzymes, antibiotics, artificial sweeteners, plant protection agents, PVC-plasticizers, and rubber antioxidants. Formic acid is used directly in dyeing, pickling, and tanning processes for leather, the coagulation of rubber, and in various cleaning products (BASF).

Furfural and its derivatives can be used as a replacement for crude-oil-based organics used in industry, such as flavoring compounds, furane resin, surface coatings, pharmaceuticals, mortar, polymers, adhesives, and moulds. Tetrahydrofurfuryl alcohol and tetrahydrofuran (THF) are also used to create specialty chemicals in a wide range of chemical synthesis (Win, 2011).

7. Gas chromatography methods for torrefaction gases

Various GC configurations are used for the analysis of pyrolysis gases that can also be used to analyze gases from the torrefaction processes. GC systems used for pyrolysis should be reviewed because those volatiles and permanent gases produced during pyrolysis are likewise produced in the torrefaction process. The authors feel that with less or no modification, these GC systems can be adequately adapted for torrefaction-gas analysis.

In the study of Prins et al. (2006b), the volatile products are split into a liquid and gas phase in a cold trap at –5°C. Liquid products collected in the cold trap were diluted with 2-butanol because not all the products dissolved in water. The diluted liquid was analyzed with high-performance liquid chromatography (HPLC) using a Chrompack Organic Acids column with detection based on refraction index. The composition of the non-condensable gas was analyzed with a Varian Micro GC with a Poraplot Q and a Molsieve column.

In the study of Deng et al. (2009), torrefaction products were removed from the hot zone to minimize the secondary reactions between the liquid and char and to maximize the solid yield. A two-necked flask immersed in liquid nitrogen was used as a trapping system for the condensable gases. Non-condensable gases went through a filter to remove the carbon soot before entering an infrared gas analysis. The gas composition and concentration were recorded continuously throughout the heating process. Finally, the weight of the biochar and the amount of liquids obtained were measured.

In Bridgeman's (2008) study, a Nicolet Magma-IR Auxiliary Equipment Module (AEM), connected to a Stanton Redcroft Simultaneous Analyser STA-780 Series, was used to perform torrefaction experiments at laboratory scale while simultaneously analyzing the volatile pyrolysis products for corresponding thermogravemetric analyzer (TGA) data.

Bergman et al. (2005) used two impinger bottles immersed in a cold bath (<5°C) for collection of the condensables during the torrefaction process. Torrefaction gas is sucked through the bottles at a pre-specified rate so that yields can be determined. A polar solvent (water or ethanol) is used in the bottles. Tests showed that two bottles removed all the condensables from the gas phase. Some aerosols are formed by quenching (going from ~200°C to ~0°C). Quantification of water was found to be unreliable due to aerosol formation and because reaction water mixes with physically bound water. It is unclear whether lipids are collected efficiently because the components of the lipid subgroup were not quantified afterwards. Analysis of the product was done with a combination of GC-FID, GC-MS, and ion chromatography. These quantified products are shown in Table 4.

2-furaldehyde (furfural)	1-Hydroxy-2-butanone
5-methyl-2-furaldehyde	Eugenol
Acetaldehyde	Isoeuganol
2(5H)-Furanone	Propionic acid
Ethylene-glycol diacetate	3,4,5 trimethoxytoluene
Ethanol	Furan-2-methanol
Pyrrole	Furan-3-methanol
Methanol	Methylacetate
Methylformiate	Hydroacetone
Cyclohexanone	Butanone
Propanal	2-butenal
Pyrrole-2-carboxaldehyde	Phenol
2-methoxyphenol	3-methoxyphenol
4-methoxyphenol	2,6 dimethoxyphenol
Acetone	3-methoxypyridine
4-methoxypyridine	2-methoxypyridine
Acetic acid	Formic acid

Table 4. Quantified condensable products produced during torrefaction of wood

Permanent gases, as shown in Table 5, are also measured using an online measurement procedure. This comprises micro-GC, by which the composition of the dry gas is determined. Before the gas is charged to the micro-GC, water is removed in a cold trap because the presence of the water disturbs the measurement.

O_2	C_2H_2
Ar	C_2H_4
CO_2	C_2H_6
CO	Benzene
N_2	Toluene
H_2	Xylenes
CH_4	

Table 5. Non-condensable gases quantified during torrefaction of wood

In Pommer et al.'s 2011 studies on torrefaction of different biomass sources using GC-MS, the most abundant compounds were identified due to the decomposition of hemicellulose, cellulose, and lignin, as shown in Table 6.

4-Allyl-syringol	Formaldehyde
Acetaldehyde	Formic acid
Acetic acid	Furaldehyde
Carbon monoxide	(3H)-Furan-2-one
Carbon dioxide	2-Furfuralcohol
Conifer alcohol	Hydroxyacetaldehyde
Eugenol	Methanol
Propanal-2-one	Sinapaldehyde
Pyranone	Synapyl alcohol
Syringaldehyde	Vanillin
4-Vinyl-syringol	Water

Table 6. GC-MS analysis of abundant compounds produced due to the torrefaction of biomass

In addition, the analysis of biomass tar can also be used for reference. Two methods for the sampling and analysis of tar produced from wood pyrolysis were compared in the study of Dufour et al. (2007). A Clarus 500 GC system (Perkin-Elmer) coupled to a Clarus 500 MS quadrupole MS system (Perkin-Elmer) was used for the analysis. The gas chromatograph was equipped with an electronically controlled split/splitless injection port. GC was carried out on a 5% diphenyl-/95% dimethylpolysiloxane fused-silica capillary column (Elite-5 ms, 60 m × 0.25 mm, 0.25 µm film thickness; Perkin-Elmer). Helium (Alphagaz 2, Air Liquide) was used as the carrier gas, with a constant flow of 1.2 mL/min. The first method used a conventional cold-trapping technique in solvent-filled impingers followed by liquid injection. The second employed a new application of multi-bed solid-phase adsorbent (SPA) tubes, followed by thermal desorption (TD). Both methods are based on GC coupled with MS. Quantification was performed with a well reproducible GC–MS method with three internal deuterated standards. The main compounds sampled by impingers with methanol or 2-propanol and SPA tubes are listed in Table 7.

Barrefors and Petersson (1995) used GC and GC–MS for volatiles produced during the pyrolysis of wheat straw. Mahinpey et al. (2009) have carried out GC and GC–MS analysis of bio-oil, biogas, and biochar from pressurized pyrolysis of wheat straw using a tubular reactor. The GC system used was a gas chromatograph (Agilent Technologies 6890 N) equipped with two columns (Porapak; molecular sieve), flame ionization detection (FID), and thermal conductivity detection (TCD). Standard gas mixtures were used for quantitative calibration, and argon was used as the carrier gas.

Benzene	1,2,4-Trimethylbenzene; 1-methylethylbenzene
Acetic acid	3,3-Dimethylphenol
Pyridine	Phenol
Pyrrole	2- or 3-Methylstyrene
Toluene	Benzofuran
Ethylbenzene	Indene
1,3-Dimethylbenzene	2- or 3- or 4-Methylphenol
Ethynylbenzene	5-Methylpyrimidine
o-Xylene	2- or 3- or 4-Methylphenol
1- or 2-Propenylbenzene	7- or 2-Methyl-benzofuran; 3-phenyl-2-propenal
7- or 2-Methyl-benzofuran	3-Phenylfuran; 1- or 2-naphthol
4,7-Dimethylbenzofuran; 1-hydroxy-5,8-dihydronaphthalene	1- or 2-Naphthol
2-Methylnaphthalene	7- or 2-Methyl-benzofuran or 3-phenyl-2-propenal
1-Methylnaphthalene	Diethenylbenzene
Fluorene	3-Phenyl-1,2-butadiene; 1-, 2-methylindene
4-Methyldibenzofuran; 9H-xanthene	3-Phenyl-1,2-butadiene; 1-, 2-methylindene
1H-phenalene	Biphenyl
2-, 3-, or 4-Methylbiphenyl	Acenaphthene
1-Phenylmethylene-1H-indene; 1,9-dihydropyrene	2-Ethenylnaphthalene
Phenanthrene	Acenaphthylene
Anthracene	2-Methylanthracene
1-Phenylmethylene-1H-indene; 1,9-dihydropyrene	6H-Cyclobuta[J,K]phenanthrene; 8,9-dihydrocyclopenta [D,E,F]phenanthrene
Dibenzofuran	1- or 2-Phenylnaphthalene
Acenaphthenone	Fluoranthene
1H phenalene	11H-benzo[A] or [B]fluorene
1-Methylpyrene	11H-benzo[A] or [B]fluorene
Pyrene	

Table 7. Main identified compounds sampled with SPA tube

The GC–MS system consisted of an Agilent HP6890GC coupled to a quadrupole mass spectrometer (Agilent 5973 Network MSD). Electron ionization (EI) was used with an ion source temperature of 230°C and an interface temperature of 280°C, with EI spectra obtained at 70 eV. In EI, the instrument was used in SCAN mode initially to confirm the identity of the 18 compounds and then in selected ion monitoring (SIM) mode for quantitative analysis. The GC system was equipped with a split/splitless inlet with a splitless sleeve containing carbofrit (4 mm i.d., 6.5 × 78.5 mm, Restek). The injector temperature was 225°C. A LEAP Technologies autosampler with a 10 μL syringe was used for injections of 1 μL at a rate of 10 μL s⁻¹. The analytical column was DB-5 ms, 5% polydiphenyl-95% dimethylpolysiloxane, 30

× 0.25 mm i.d., and 0.25 μm film thickness (J&W Scientific). The carrier gas was helium (UHP) at a constant flow of 1.0 mL min^{-1}. The oven temperature program had an initial temperature of 65°C, which was held for 4 min, rising by 10°C/min to 120°C, 15°C/min to 170°C, 3°C/min to 200°C, and finally 30°C/min to 300°C, where it was held for 5 min for a total run time of 31.17 minutes. This temperature program was selected to provide adequate separation of most of the 18 compounds of interest, as shown in the total ion chromatogram (TIC) for a standard and sample. Only two compounds coeluted (pyrocatechol and phenol 2-methoxy 4-methyl), and the remaining compounds are provided in Table 8.

Phenol	4-methylcatechol	3,5-dimethoxy-4'-hydroxy acetophenone
Methylhydroquinone	2,6-dimethoxyphenol	Syringaldehyde
Guajacol	Eugenol	2-methoxy -4-propylphenol
4-ethyl phenol	Vanillin	phenol 2 -methoxy 4-methyl
pyrocatechol	Isoeugenol	3-methoxycatechol
4'-hydroxy-3'-methoxyacetophenone	4-ethyl guaiacol	2-propanone, 1-(4-hydroxy-3-methoxyphenol

Table 8. GC and GC–MS analysis of gases produced during pyrolysis of wheat straw

In the studies conducted by Bacidan et al. (2007), product distribution and gas release in pyrolysis of thermally thick biomass residues samples have used micro GC and FTIR analyzer for gases like CO_2, CO, CH_4, H_2, C_2H_2, C_2H_6, and C_2H_4. The FTIR analysis of the gases was performed with a Bomem 9100 analyser (sampling line and cell heated at 176 °C with a volume of 5l and an optical path length of 6.4 m). The instrument is equipped with a DTGS detector at the maximum resolution of 1 cm^{-1}. The gas samples were also quantified online using a Varian CP-4900 micro-GC equipped with two TCD detectors and a double injector connected to two columns: (1) a CP-PoraPLOT Q column (10 m length, 0.25 mm inner diameter and 10 μm film thickness produced by Varian, Inc.) to separate and quantify CO_2, CH_4, C_2H_2 + C_2H_4 (not separated), and C_2H_6 ; and (2) a CP-MolSieve 5 Å PLOT column (20 m length, 0.25 mm inner diameter and 30 μm film thickness produced by Varian, Inc.) to analyse H_2, O_2, CH_4, CO, and N_2. Helium and argon were used, respectively, as carrier gases in the two columns.

Lappas et al. (2002) focused on identifying the composition of the gases produced during biomass pyrolysis, where an online connected GC (HP 5890) with two detectors (TCD and FID) was used, along with a split/splitless injector and four columns (OV 101, Molecular sieve 5A, Porapaq N, and a capillary Poraplot Al2O3/KCl). The composition of the regenerator gas exit stream (i.e., flue gases) was determined with a GC (HP 5890) detector (TCD, Molecular sieve 5A and Porapaq N columns), and the hydrocarbons, CO and CO_2 were analyzed.

Qing et al. (2011) analyzed the gases and volatiles produced during the pyrolysis of wheat straw. They used a Shimazu QP2010 Plus GC-MS, and the GC column type was a DB-Wax fused silica capillary column (30 m × 0.25 mm; i.e., film thickness = 0.25 μm).

And finally, Wang et al. (2011) concentrated on the pyrolysis of pine sawdust by using a GC-MS (Finnigan Trace MS). The column used in these studies was a GC-MS with a capillary column DB-1301 (30 m x 0.25mm; i.e., film thickness = 0.25 μm). Helium was the carrier gas, with a constant flow of 0.5 ml/min. The MS was operated in electron ionization mode with a 70 eV ionization potential, and an m/z range from 30 to 500 was scanned. The peaks were identified based on the computer matching of the mass spectra with the National Institute of Standards and Technology (NIST) library. The summary of the systems used for analysis of torrefaction and pyrolysis gas analysis is given in Table 9.

S. No	GC configuration	Process	Volatiles and permanent gases		Reference
			Volatiles	Gases	
1	HPLC using Chrompack organic acids column for volatiles. Varian micro gas chromatography with Porapolt Q and Molsieve column for non condensable gases	Torrefaction	Water, acetic, lactic acid, formic acid, furfural, hydroxyl acetone, phenol, methonal	CO, CO_2	Prins et al. (2006b)
2	Infrared gas analysis (Gasboard-5110, Wuhan, China).	Torrefaction	Moisture, acetic acid and other oxygenates	CO_2, CO and small amounts of methane	Deng et al. (2009)
3	TGA-FTIR (A Nicolet Magma-IR AEM connected to a Stanton Redcroft Simultaneous Analyser STA-780 Series)	Torrrefaction	Volatiles	CO_2, CO	Bridgeman et al. (2008)
5	GC–MS (A Clarus 500 GC system (Perkin-Elmer, Shelton, CT, USA) coupled to a Clarus 500 MS quadrupole MS system (Perkin-Elmer)	Torrefaction	Benzene, acetic acid, pyridine, pyrole, toluene, ethylbenzene, 1,3-dimethylbenzene,etc		Dufour et al. (2007)
6	GC-FID and GC-MS	Torrefaction	2-furaldehyde (furfural), 1-Hydroxy-2-butanone, acetic acid, formic acid etc.	O_2, Ar, CO_2, H_2, CO	Bergman et al. (2005)
7	GC-MS	Torrefaction	Acetic acid, formic acid, formaldehyde, vanillin, methanol, acetaldehyde etc.	CO, CO_2	Pommer et al. (2011)

S. No	GC configuration	Process	Volatiles and permanent gases		Reference
8	GC-MS	Pyrolysis	Acetic acid, formic acid, formaldehyde etc.		Barrefors and Petersson (1995)
9	GC-MS (Agilent HP6890GC coupled to a quadrupole mass spectrometer) (Agilent 5973 Network MSD) GC (Agilent Technologies 6890 N) equipped with two columns (Porapak; Molecular sieve), flame ionization detection (FID), and thermal conductivity detection (TCD)	Pyrolysis	Phenol, vanillin, acetophenone, syringaldehyde etc.	CO, CO_2 etc	Mahinpey et al. (2009)
10	FTIR (Bomem 9100 analyser) and Micro GC	Pyrolysis	Hydrocarbons	CO, CO_2	Bacidan et al. (2007)
11	Gas chromatograph-GC (HP 5890) with two detectors (TCD and FID)	Pyrolysis		CO, CO_2	Lappas et al. (2002)
12	Shimazu QP2010 Plus GC-MS. GC Column Type: A DB-Wax fused silica capillary column (30 m × 0.25 mm, i.e., film thickness = 0.25 μm)	Pyrolysis			Qing and Wu (2011)
13	GC-MS: The column used in GC-MS was a capillary column DB-1301 (30mx0.25mm i.d., 0.25μm film thickness) and MS was operated in electron ionization mode with a 70 eV ionization potential.	Pyrolysis	Higher hydrocarbons	CO, CO_2	Wang et al. (2011)

Table 9. GC systems for torrefaction and pyrolysis gas analysis

8. Conclusions

Torrefaction of biomass produces a product that has a higher energy content compared to that of raw biomass. During torrefaction, biomass loses moisture as well as some condensable and non-condensable gases. The condensable gases include water, organics, and lipids. The liquid contains the free and bound water that has been released from the biomass by evaporation. The organics subgroup (in liquid form) consists of organics that are mainly produced during devolatilization, depolymerization, and carbonization reactions in the biomass. The knowledge of the composition of volatiles produced in the temperature range of torrefaction is a topic of interest as it helps in the raw material adaption process, process control, process behavior and operation, energy process and energy optimization, and in the production of green chemicals. Lipids are a group of compounds that are present in the original biomass and contain compounds such as waxes and fatty acids. The non-condensable gases are CO_2, CO, and small amounts of methane. GC configuration plays an important role in the accurate identification of the compounds. A combination of different detectors, such as thermal, flame, and photo ionization detectors, in combination with a mass spectrometer, are used for the profiling of both condensable and non-condensable gases. It can be concluded that a GC-MS can help to analyze most of the volatiles emitted from biomass during torrefaction as well as a micro GC for the analysis of permanent gases like CO, CO_2, and CH_4.

9. Acknowledgements

The authors acknowledge Leslie Park Ovard for her valuable contribution to the manuscript and Barney Hadden, Ph.D, Gordon Holt, Lisa Plaster, and Allen Haroldsen from INL's R&D Publication Support Service for editorial and graphics support. This work is supported by the U.S. Department of Energy, under DOE Idaho Operations Office Contract DE-AC07-05ID14517. Accordingly, the U. S. Government retains and the publisher, by accepting the article for publication, acknowledges that the U. S. Government retains a nonexclusive, paid-up, irrevocable, world-wide license to publish or reproduce the published form of this manuscript, or allow others to do so, for U. S. Government purposes.

10. Disclaimer

11. References

Arias, B.R., Pevida, C.G., Fermoso, J.D., Plaza, M.G., Rubiera, F.G., and Pis Martinez, J.J. (2008). Influence of Torrefaction on the Grindability and Reactivity of Woody

Biomass, *Fuel Processing Technology*, Vol.89, No.2, (February 2008), pp. 169–175, ISSN 0378-3820

Barrefors, G., and Petersson, G. (1995). Assessment by gc and gc-ms of Volatile, *Journal of Chromatography A*, Vol.710, No.1, (1 September 1995), pp. 71-77, ISSN 0021-9673

BASF: The Chemical Company, Formic Acid (webpage).
http://www2.basf.us/specialtyintermediates/formic_acid.html

Becidan, M., Skreiberg, Ø., and Hustad, J.E. (2007). Products Distribution and Gas Release in Pyrolysis of Thermally Thick Biomass Residues Samples, *Journal of Analytical and Applied Pyrolysis*, Vol.78, No.1, (January 2007), pp. 207–213, ISSN 0165-2370

Bergman, P.C.A., and Keil, J.H.A. (2005). Torrefaction for Biomass Upgrading, *14th European Biomass Conference & Exhibition*, Paris, France, 17–21 October 2005

Bergman P.C.A., Boersma, A.R., Kiel, J.H.A., Prins, M.J., Ptasinski, K.J., and Janssen, F.J.J.G. (2005a). Torrefied Biomass for Entrained-flow Gasification of Biomass, ECN Report #ECN-C--05-026

Bergman, P.C.A., Boersma, A.R., Zwart, R.W.R., and Kiel, J.H.A. (2005). Torrefaction for Biomass Co-firing in Existing Coal-fired Power Stations: BIOCOAL, ECN Report #ECN-C-05-013

Bosboom, J. C., Janssen, H. G., Mol, H. G. J., and Cameras, C. A. (1996). Large-volume Injection in Capillary Gas Chromatography using a Programmed-temperature Vaporizing Injector in the On-column or Solvent-vent Injection Model, *Journal of Chromotography A*, Vol.724, No.1-2, (16 February 1996), pp. 384-391, ISSN 0021-9673

Bourgois, J.P., and Guyonnet, R. (1988). Characterization and Analysis of Torrefied Wood, *Wood Science and Technology*, Vol.22, No.2, (June 1988), pp. 143–155, ISSN 0043-7719

Bourgois, J.P., and Doat, J. (1984). Torrefied Wood from Temperate and Tropical Species: Advantages and Prospects, In: *Bioenergy 84, Proceedings of the International Conference on Bioenergy, Goteborg, Sweden, 15–21 June, 1984*, H. Egneus, A. Ellegard, P. O'Keefe, and L. Kristofferson (Eds.), Elsevier Applied Science Publishers, pp. 153–159, ISBN 978-085-3343-51-6

Bridgeman, T.G., Jones, J.M., Shield, I., and Williams, P.T. (2008). Torrefaction of Reed Canary Grass, Wheat Straw, and Willow to Enhance Solid Fuel Qualities and Combustion Properties, *Fuel*, Vol.87, No.6, (May 2008), pp. 844–856, ISSN 0016-2361

Demirbas, A. (2009). Pyrolysis Mechanisms of Biomass Materials, *Energy Sources, Part A: Recovery, Utilization, and Environmental Effects*, Vol.31, No.13, (13 May 2009), pp. 1186–1193, ISSN 1556-7230

Deng, J., Wang, G., Kuang, J., Zhang, Y., and Luo, Y. (2009). Pretreatment of Agricultural Residues for Co-Gasification Via Torrefaction, *Journal of Analytical and Applied Pyrolysis*, Vol.86, No.2, (November 2009), pp. 331–337, ISSN 0165-2370

Dufour A., Girods, P., Masson, E., Normand, S., Rogaume, Y., and Zoulalian, A. (2007). Comparison of Two Methods of Measuring Wood Pyrolysis Tar, *Journal of Chromatography A*, Vol.1164, No.1-2 (14 September 2007), pp. 240–247, ISSN 0021-9673

Fonseca Felfli, F., Luengo, C.A., Bezzon, G., and Beaton Soler, P. (1998). Bench Unit for Biomass Residues Torrefaction, *Biomass for Energy and Industry, Proceedings of the International Conference, Würzburg, Germany, 8-11 June 1998,* H. Kopetz, T. Weber, W. Palz, P. Chartier, and G.L. Ferrero (Eds.), C.A.R.M.E.N., Rimpar, Germany, pp. 1593–1595

Goorts, M.P. (2008). Applying Gas Chromatography to Analyze the Composition and Tar Content of the Product Gas of a Biomass Gasifier, Eindhoven University of Technology, Master's Thesis, Report Number WVT 2008.24, Available from http://w3.wtb.tue.nl/fileadmin/wtb/ct-pdfs/Master_Theses/Final_report_martijn_goorts.pdf

Grob, R.L., and Barry, E.F. (2004). *Modern Practice of Gas Chromatography (4th Ed.),* John Wiley & Sons, Inc., Hoboken, NJ, ISBN 0-471-22983-0

Guiochon, G., and Guillemin, C.L. (1988). *Quantitative Gas Chromatography for Laboratory Analyses and On-Line Process Control, (Journal of Chromatography Library,* No.42), Elsevier Ltd., Amsterdam, the Netherlands, ISBN 978-044-4428-57-8

Hübschmann, H-J. (2008). *Handbook of GC/MS: Fundamentals and Applications,* Wiley-VCH, Weinheim, Germany, ISBN 3-527-31427-X

ICIS, Acetic Acid Uses and Market Data (webpage).
http://www.icis.com/v2/chemicals/9074779/acetic-acid/uses.html

ICIS, Methanol Uses and Market Data (webpage).
http://www.icis.com/v2/chemicals/9076035/methanol/uses.html

IEA Bioenergy, Task 34 – Pyrolysis of Biomass, Leader: Doug Elliott (webpage).
http://www.pyne.co.uk/

Kaufmann, A. (1997). Determination of polar analytes in aqueous matrices by purge and trap, *Journal of High Resolution Chromatography,* Vol.20, No.1, (January 1997), pp. 10–16, ISSN 0935-6304

Lappas, A., Samolada, M., Latridis, D., Voutetakis, S., and Vasalos, I. (2002). Biomass Pyrolysis in a Circulating Fluid Bed Reactor for the Production of Fuels and Chemicals, *Fuel,* Vol.81, No.16, (1 November 2002), pp. 2087-2095, ISSN 0016-2361

Mahinpey, N., Murugan, P., Mani, T., and Raina, R. (2009). Analysis of Bio-Oil, Biogas, and Biochar from Pressurized Pyrolysis of Wheat Straw using a Tubular Reactor, *Energy & Fuels,* Vol.23, No.5, (21 May 2009), pp. 2736-2742, ISSN 0887-0624

Mohan, D., Pittman, C.U., and Steele, P.H. (2006). Pyrolysis of Wood/Biomass for Bio-Oil: A Critical Review, *Energy & Fuels,* Vol.20, No.3, (May 2006), pp. 848–889, ISSN 0887-0624

Pach, M., Zanzi, R., and Björnbom, E. (2002). Torrefied Biomass a Substitute for Wood and Charcoal, In: *Proceedings of the 6th Asia-Pacific International Symposium on Combustion and Energy Utilization, Kuala Lumpur, Malaysia, May 20–22, 2002.*

Pavia, D.L., Lampman, G.M., Kritz, G.S., and Engel, R.G. (2006). *Introduction to Organic Laboratory Techniques: A Microscale Approach (4th Ed.),* Cengage Learning, Brooks/Cole Publishing Co., St. Paul, MN, (February 2006), pp. 797–817, ISBN 978-049-5016-30-4

Pentananunt R., Mizanur Rahman, A.N.M., and Bhattacharya, S.C. (1990). Upgrading of Biomass by Means of Torrefaction, *Energy*, Vol.15, No.12, (December 1990), pp. 1175-1179, ISSN 0360-5442

Pommer, L., Gerber, L., Olofsson, I., Wiklund-lindstrom, S., and Nordin, A. (2011). Gas composition from biomass torrefaction-preliminary results, *19th European Biomass Conference and Exhibition, Berlin, Germany, 6-10 June 2011*, Available from http://www.conference-biomass.com/fileadmin/DATA/Documents_and_ Pictures/PAST_CONFERENCES/Previous_Conferences/18th_EU_BC_E_Proceedi ngs/TOC_Biomass_2010.pdf

Poy, F.S., Visani, F., and Terrosi, F. (1981). Automatic Injection in High Resolution Gas Chromatography: A Programmed Temperature Vaporizer as a General Purpose Injection System, *Journal of Chromatography A*, Vol.217, No.1, (6 November 1981), pp. 81-90, ISSN 0021-9673

Prins, M.J., Ptasinski, K.J., and Janssen, F.J.J.G. (2006). Torrefaction of Wood: Part 1 - Weight Loss Kinetics, *Journal of Analytical and Applied Pyrolysis*, Vol.77, No.1, (August 2006), pp. 28-34, ISSN 0165-2370

Prins, M.J., Ptasinski, K.J., and Janssen, F.J.J.G. (2006a). More Efficient Biomass Gasification via Torrefaction, *Energy*, Vol.31, No.15, (December 2006), pp. 3458-3470, ISSN 0360-5442

Prins, M.J., Ptasinski, K.J., and Janssen, F.J.J.G. (2006b). Torrefaction of Wood: Part 2 - Analysis of Products, *Journal of Analytical and Applied Pyrolysis*, Vol.77, No.1, (August 2006), pp. 35-40, ISSN 0165-2370

Punrattanasi, W and Spad, C. (2011). Environmental Sampling and Monitoring Primer: Gas Chromatography (webpage), Virginia Polytechnic Institute and State University (Virginia Tech) Department of Civil and Environmental Engineering, Available from http://www.cee.vt.edu/ewr/environmental/teach/smprimer/gc/gc.html

Sadaka, S., and Negi, S. (2009). Improvements of Biomass Physical and Thermochemical Characteristics via Torrefaction Process, *Environmental Progress & Sustainable Energy*, Vol.28, No.3, (October 2009), pp. 427-434, ISSN 1944-7442

Scott, R.P.W. (2003). *Chrom-Ed Book Series: Book 4 - Gas Chromatography Detectors* (e-book), LibraryForScience, LLC, Available from http://www.chromatography-online.org/2/contents.html

SHU, Gas Chromatography (webpage). http://teaching.shu.ac.uk/hwb/chemistry/tutorials/chrom/gaschrm.htm

Tumuluru, J.S., Sokhansanj, S., Hess, R.J., Wright, C.T. and Boardman, R.D. (2011). A review on biomass torrefaction process and product properties for energy applications, *Industrial Biotechnology*, Vol. 7, No. 5, (October, 2011), pp.384-401, ISSN: 1550-9087

United Nations (UN). (1998). Kyoto Protocol to the United Nations Framework Convention on Climate Change, Available from http://unfccc.int/resource/docs/convkp/kpeng.pdf

White, R.H., and Dietenberger, M.A. (2001). Wood Products: Thermal Degradation and Fire, *The Encyclopedia of Materials: Science and Technology*, KH Jürgen Buschow, RW Cahn,

MC Flemings, B Ilschner, EJ Kramer, S Mahajan, and P Veyssière (Eds.), Elsevier, Ltd., Amsterdam, the Netherlands, pp. 9712–9716, ISBN 978-008-0431-52-9

Win, D.T. (2005). Furfural-Gold from Garbage, *AU Journal of Technology*, Vol.8, No.4, (Apr. 2005), pp. 185-190, Available from
http://www.journal.au.edu/au_techno/2005/apr05/vol8no4_abstract04.pdf

Zanzi, R, Ferro, D.T., Torres, A., Soler, P.B., and Bjornbom, E. (2002). Biomass Torrefaction, In: *Proceedings of the 6th Asia-Pacific International Symposium on Combustion and Energy Utilization, Kuala Lumpur, Malaysia, May 20–22, 2002.*

4

Application of Gas Chromatography to Exuded Organic Matter Derived from Macroalgae

Shigeki Wada and Takeo Hama

Shimoda Marine Research Center, University of Tsukuba,
Life and Environmental Sciences, University of Tsukuba,
Japan

1. Introduction

Macroalgae are the most important primary producers in coastal environments, because their productivity per square community area is comparable to that of tropical rain forests (Mann, 1973; Yokohama et al., 1987). Although a half of the products constitute the organic matter of algal body (Hatcher et al., 1977), considerable part of them would be exuded and dissolved into ambient seawater (Khailov & Burlakova, 1969; Sieburth, 1969; Hatcher et al., 1977; Abdullah & Fredriksen, 2004; Wada et al., 2007). The organic materials released into ambient seawater induces some alteration of marine and atmospheric environments. For example, some volatile compounds released from macroalgae would escape into air, and play as ozone-depleting substances (Lovecock, 1975; Laturnus et al., 2010). Phenolic compounds are known as a kind of exudates, which acts as a form of defense to the algal body from herbivores (van Alstyne, 1988). A part of the phenolic compounds are likely to dissolve into seawater, and the light shielding role as a component of colored organic matter was suggested in coastal environments (Wada et al., 2007). In addition to these compounds, carbohydrates such as mucopolysaccharides are important component of released organic matter on the body surface, and a part of them is considered to be released into seawater (Wada et al., 2007). It is known that bacterial community acting on the macroalgae is variable depending on the carbohydrate composition of exudates (Bengtsson et al., 2011), and the bacterial community structure probably changes the carbon flux around algal bed. The characterization of organic compounds derived from macroalgae is necessary to elucidate the biogeochemical role of macroalgae, because the effect to biogeochemical processes is variable depending on each organic compound. Thus, quantitative and qualitative approaches have been carried out so far. Application of gas chromatography (GC) would be effective for some of these exuded compounds. In this review, we focused on the analytical methods using GC for analysis of compounds originated from macroalgae.

2. Volatile halogenated organic matter

2.1 Macroalgal release of volatile halogenated organic matter

Depletion of ozone in stratosphere has been focused as one of the most serious global environmental issues, and reaction of halogenated compounds with ozone has been recognized as the destruction mechanisms of ozone (Farman et al., 1985; Crutzen & Arnold, 1986; Solomon, 1990; Anderson et al., 1991). Such volatile halogenated compounds are not only originated from anthropogenic sources, but also biogenic sources such as macroalgae. Due to such interests, volatile halogenated organic compounds (VHOC) have been intensely studied so far, and development of analytical technique of macroalgal VHOC has been recognized.

2.2 Pre-treatment before injection

When analysis of macroalgal VHOC in seawater samples are carried out, pre-treatment of the sample is important procedure for achievement of high recovery yield. There are several kinds of methods such as closed-loop stripping (CLSP), headspace, liquid-liquid extraction, purge-and-trap (P&T) and solid-phase micro extraction (SPME) methods (Table 1). Within these methods, CLSP had been applied in earlier year (Gschwend et al., 1985; Newman and Gschwend, 1987), but there is a problem that the CLSP method is inappropriate for analyses of extremely volatile compounds (e.g., CH_3Cl and CH_3Br) or relatively involatile compounds (e.g., CBr_4 and CHI_3) (Gschwend et al., 1985). In headspace method, seawater samples were brought into equilibrium with a gas (usually nitrogen) (Lovelock 1975; Manley & Dastoor, 1987; Manley & Dastoor, 1988; Manley et al., 1992). In the case that the concentrations of target compounds were low, cryo-concentration had been also applied (Manley & Dastoor, 1987; Manley & Dastoor, 1988; Manley et al., 1992). Liquid-liquid extraction is that solvent containing internal standard is added into seawater sample, and a part of the solvent phase was injected to GC. This method allows us to make analyses with simple instrumental set-up and the total analysis time is short (Abrahamsson & Klick, 1990; Laturnus et al., 1996; Manley & Barbero, 2001). P&T method have been the most widely-applied method in recent years. Sample seawater is purged with nitrogen or helium gases, and the target compounds in the gas phase were concentrated with cold trap (sometimes with adsorbent; Ekdahl & Abrahamsson, 1997). After degassing, the trapped compounds were transferred into GC instrument by heating (Schall et al., 1994; Nightingale et al., 1995; Laturnus et al., 2004; Weinberger et al., 2007; Laturnus et al., 2010). Recently, SPME technique has been also applied for VHOC originated from macroalgae. SPME fibre is used for trapping of VHOC after purging with pure nitrogen. In case of determination of VHOC in seawater spiked by standards, quantitative quality was confirmed at the concentration level around 100 ng l^{-1} of VHOC (Bravo-Linares et al., 2010).

2.3 GC Instruments

Analyses of extracted VHOC compounds have been mainly carried out by gas chromatography with electron capture detector (GC-ECD) (e.g., Manley & Dastoor, 1987; 1988; Schall et al., 1994; Laturnus et al., 1996; Manley & Barbero, 2001; Laturnus et al., 2004) or gas chromatography/mass spectrometry (GCMS) (e.g., Gschwend et al., 1985; Newman & Gschwend, 1987; Marshall et al., 1999; Bravo-Linares et al., 2010). Although GC-ECD has been commonly applied, it is necessary to confirm the retention time of each compound using authentic standards (Giese et al., 1999) or GCMS (Manley and Dastoor, 1987) to identify each component. On the other hand, application of GCMS has an advantage that it is reliable to characterize the compounds based on the mass spectrum. Capillary column is also an important part of GC for separation of each compound, and various kinds of columns (e.g., SE 54, BP-624, Rtx 502.2) have been applied. Most of them have mid-polarity, which have been commercially recommended for analysis of volatile organic compounds.

Pre-treatment	Target compounds	Sample	Instruments	References
Headspace	CH_3Cl, CH_3Br, CH_3I	Seawater	GC-ECD	Lovelock (1975)
Closed loop stripping	$CHBr_3$, $CHBr_2Cl$, CH_2Br_2	2 Brown algae and 2 Green algae,	GCMS	Gschwend et al. (1985)
Closed loop stripping	$(CH_3)_2CHBr$, $CH_3CH_2CH_2Br$, $CH_3(CH_2)_4Br$, CH_3I, C_2H_5I, C_3H_7I, C_4H_9I, $C_5H_{11}I$, $C_6H_{13}I$, CH_2Br_2, $CHBr_3$, CH_2I_2, $CHBr_2I$, CH_3SCH_3, CH_3SSCH_3	1 Brown algae	GCMS	Newman & Gschwend (1987)
Headspace	CH_3Cl, $CHBr$, CH_3I	1 Brown algae and seawater	GC-ECD	Manley & Dastoor (1987)
Headspace	CH_3I	5 Brown algae	GC-ECD	Manley & Dastoor (1988)
Headspace	$CHBr_3$, CH_2Br_2, CH_3I	6 Brown algae, 3 Red algae, 2 Green algae and seawater	GC-ECD	Manley et al. (1992)
Headspace	$CHBr_3$, CH_2Br_2	1 Brown algae	GC-ECD	Goodwin et al. (1997)
Liquid-liquid	$CHBr_3$, CH_2Br_2, $CHBr_2Cl$, CH_2I_2, CH_2ClI, CCl_4, CH_3CCl_3, $CHClCCl_2$	Seawater	GC-ECD	Klick (1992)
Liquid-liquid	CH_2Br_2, $CHBrCl_2$, CH_2ClI, $CHBr_2Cl$, 1,2-$C_2H_4Br_2$, CH_2I_2, $CHBr_3$	9 Brown algae, 15 Red algae, 2 Green algae and 2 Crysophyta	GC-ECD	Laturnus et al. (1996)
Liquid-liquid	CH_2Br_2, $CHBrCl_2$, CH_2ClI, $CHBr_2Cl$, 1,2-$C_2H_4Br_2$, CH_2I_2, $CHBr_3$	11 Brown algae, 4 Red algae and 6 Green algae	GC-ECD	Laturnus (1996)
Liquid-liquid	CCl_4, $CHCl=CCl_2$, $CCl_2=CCl_2$, CH_2Br_2, $CHBr_3$, CH_2ClI, C_4H_9I, CH_2I_2	Seawater	GC-ECD	Abrahamsson & Ekdahl (1996)
Liquid-liquid	$CHBr_3$	1 Green algae	GC-ECD	Manley & Barbero (2001)

Pre-treatment	Target compounds	Sample	Instruments	References
Purge and trap	CH_3I, $CHCl_3$, CH_3CCl_3, CCl_4, CH_2Br_2, $CHBrCl_2$, CCl_2CCl_2, $CHBr_2Cl$	5 Brown algae, 3 Red algae, 3 Green algae and seawater	GC-ECD	Nightingale et al. (1995)
Purge and trap	CH_3I, CH_2I_2, CH_2ClI, $CH_3CH_2CH_2I$, CH_3CHICH_3, CH_2Br_2, $CHBr_3$, $CHBrCl_2$, $CHBr_2Cl$	3 Brown algae	GC-ECD	Schall et al. (1994)
Purge and trap	C_2H_5Br, $1,2-C_2H_4Br_2$, C_2H_5I, CH_2ClI, CH_2I_2, CH_3Br, CH_2BrCl, CH_2Br_2, $CHBrCl_2$, $CHBr_2Cl$, $CHBr_3$	4 Brown algae, 2 Red algae and 4 Green algae	GC-ECD	Laturnus (1995)
Purge and trap	C_2HCl_3, C_2Cl_4	5 Brown algae, 17 Red algae and 6 Green algae	GC-ECD	Abrahamsson et al. (1995)
Purge and trap	$CHBr_3$, $CHCl_3$, CH_3CCl_3, CH_2Br_2, CH_2I_2, CH_3I, $CHBr_2Cl$, $CHCl_2Br$, CH_2ClI, CCl_4, $Cl_2C=CCl_2$, $Cl_2C=CHCl$, C_4H_9I, sec-C_4H_9I, C_3H_7I	1 Green algae	GC-ECD	Mtolera et al. (1996)
Purge and trap	CH_3I, CH_2I_2, C_4H_9I, CH_2ClI, $CHBr_3$, $CHBr_2Cl$, $CHBrCl_2$, $CHCl_3$, $HCCl=CCl_2$, CH_2Br_2	2 Red algae and 1 Green algae	GC-ECD	Pedersen et al. (1996)
Purge and trap	CH_2I_2, CH_2BrCl. $CHBrCl_2$, C_2H_5Br,. $1,2-C_2H_4Br_2$, C_2H_5I, CH_2ClI	3 Red algae and 1 Green algae	GC-ECD	Laturnus et al. (1998)
Purge and trap	CH_3I, CH_2I_2, CH_2ClI, $CHBr_3$, CH_2Br_2, CH_3CH_2I, $CH_3CH_2CH_2I$, CH_3CHICH_3, $CH_3CH_2CH_2CH_2I$, $CH_3CH(CH_3)CH_2I$, $CH_3CH_2CHICH_3$	2 Brown algae, 19 Red algae and 3 Green algae	GC-ECD	Giese et al. (1999)
Purge and trap	$CHBr_3$, $C_2H_2Br_2$, CH_2Br_2	1 Red algae	GCMS	Marshall et al. (1999)

Pre-treatment	Target compounds	Sample	Instruments	References
Purge and trap	C_2H_5I, 2-C_3H_7I, 1-C_3H_7I, CH_2ClI, 2-n-C_4H_9I, CH_2I_2, $CHBr_3$	1 Red algae	GC-ECD	Laturnus et al. (2000)
Purge and trap	C_2HCl_3, C_2Cl_4	1 Red algae	GCMS	Marshall et al. (2000)
Purge and trap	CH_3Br, CH_3Cl, CH_2Cl_2, $CHCl_3$, CH_3I, CH_2ClI, CH_3CH_2I	2 Brown algae, 1 Red algae and 3 Green algae	GC-ECD	Baker et al. (2001)
Purge and trap	CH_3Cl, CH_3Br, CH_2BrCl, CH_2Br_2, $CHBrCl_2$, $CHBr_2Cl$, $CHBr_3$, 1,2-$C_2H_4Br_2$, CH_3I, CH_2ClI, CH_2I_2, C_2H_5I, 1-C_3H_7I, 2-C_3H_7I, 1,3-C_3H_6ClI, 1-n-C_4H_9I, 20n-C_4H_9I, 1-iso-C_4H_9I	11 Brown algae, 11 Red algae and 8 Green algae	HRGC-ECD/MIP AED	Laturnus (2001)
Purge and trap	$CHCl_3$, C_2HCl_3, C_2Cl_4, $CHBr_3$, CH_2Br_2, $CHClBr_2$, CH_2BrCl, $CHCl_2Br$, CH_2I_2, CH_3I, C_2H_5I, CH_2ClI, C_3H_7I, iso-C_3H_7I, C_4H_9I, sec-C_4H_9I	1 Brown algae, 4 Green algae and 1 Diatom	GC-ECD	Abrahamsson et al. (2003)
Purge and trap	C_2Cl_4, C_2HCl_3, CCl_4, $CHCl_3$, CH_2Cl_2, CH_3CCl_3, CH_3I, CH_2Br_2, $CHBr_3$, $CHBrCl_2$	3 Brown algae, 1 Red algae and 1 Green algae	GC-ECD	Laturnus et al. (2004)
Purge and trap	CH_2ICl, CH_2I_2, CH_2IBr, C_2H_5I, 2-C_3H_7I, CH_3I, CH_2Br_2, $CHBr_3$	seawater	GCMS	Jones et al. (2009)
Purge and trap	CH_3I, CH_2ClI, CH_2I_2, CH_2Cl_2, $CHCl_3$, CH_3CCl_3, CCl_4, C_2HCl_3, C_2Cl_4, CH_2Br_2, $CHBrCl_2$, $CHBr_2Cl$, 1,2-$EtBr_2$, $CHBr_3$	1 Brown algae	Not described	Laturnus et al. (2010)
SPME	CH_3Br, CH_2I_2, $CCl_2=CCl_2$, $CHCl_2Br$	4 Brown algae, 2 Red algae and 2 Green algae	GCMS	Bravo-linares et al. (2010)

Table 1. List of studies and methodologies used on VHOC originated from macroalgae

2.4 Estimation of the importance of macroalgal VHOC

In order to evaluate macroalgal release of VHOC, an effective approach is seawater sampling in and out of algal bed in the field or incubation experiment in closed system. Analysis of seawater sample often showed higher concentration of VHOC around algal bed compared with offshore region, strongly suggesting the significant release of VHOC from macroalgae (Lovecock, 1975; Manley & Dastoor, 1987; Klick, 1992; Manley et al., 1992; Nightingale et al., 1995). The incubation experiment can provide production rate of VHOC for each macroalgal species (Gschwend et al., 1985; Manley & Dastoor, 1987). Assuming that the estimated production rate of VHOC is comparable with other macroalgae, some researchers estimated global VHOC production rates (Gschwend et al., 1985; Manley & Dastoor, 1987). Although they multiplied the production rate per algal biomass which is experimentally defined by standing crop in global ocean, the most serious problem is that no detailed investigation on algal biomass in global ocean has been reported yet. Calculation of algal biomass in global ocean was estimated based on coastal length by De Vooys (1979) who also provided an estimation of primary production in global ocean as 0.03 PgC y^{-1}. No estimate of biomass has been published by other researchers, but primary production was revised by Charpy-Roubaud & Sournia (1990). They made estimation of primary production as 2.55 PgC y^{-1} based on algal community area, which is two orders of magnitude higher than the values of De Vooys (1979). Such discrepancy with the previous estimates implies that there is no reliable value of macroalgal parameters such as biomass and productivity in global ocean. In addition to the development of analytical technique, estimates of algal biomass in global ocean will be also required for understanding the macroalgal contribution to ozone-depletion.

3. Phenolic compounds

3.1 Macroalgal phenolic compounds

Macroalgae synthesize phenolic compounds, and a part of them is likely exuded (Paul et al., 2006). Considering that the phenolic compounds accumulated in the outer cortical layer of the thalli (Shibata et al., 2004; Paul et al., 2006), these materials would be actively released. Dissolution of phenolic compounds was also supported by *in situ* field experiment (Wada et al., 2007), in which absorption spectra of macroalgal excretion were relatively similar to those of the materials containing aromatic ring (lower exponential slope of the absorption spectra) (Blough & Del Vecchio, 2002). The UV absorbing property of phenolic compounds (Łabudzińska & Gorczyńska 1995) suggests the attenuation of UV radiation to seawater by phenolic compounds originated from macroalgae. Since biological activity in surface seawater is affected by UV penetration (Blough & Del Vecchio, 2002), analysis of phenolic compounds shows the interaction between macroalgae and other marine organisms.

3.2 Cupric oxide oxidation

Analytical procedure for macroalgal phenolic compounds is cupric oxide (CuO) oxidation method, in which polymeric compounds are degraded to small molecules that can be quantified by GC instrument. Although this technique has been mainly applied to lignin which also contains phenolic structure, some researchers had applied it to macroalgal materials (Goni & Hedges, 1995). Generally, the samples were reacted with CuO at 150-

170°C for 3 h under alkaline conditions, and acidified after the reaction. The oxidized fraction was extracted with ethyl ether, and the solvent was evaporated. The products were dissolved in pyridine, and they are derivatised to trimethylsilyl derivatives for gas chromatographic analysis. Molecules containing aromatic ring were identified using GC or GCMS (Goni & Hedges, 1995), but a possibility that some of them originated from protein, because 3 kinds of amino acids (phenylalanine, tryptophan and tyrosine) have aromatic group in their molecule. Since m-hydroxybenzoic acid and 3,5-dihydroxybenzoic acid has been suggested as non-amino acid derived from organic compounds containing phenolic structure such as tannin, these products are likely useful as an indicator of macroalgal materials when we use CuO oxidation method.

3.3 Application of CuO oxidation method for macroalgal exudates

There are just a few studies on analysis of phenolic compounds using CuO oxidation method (Goni & Hedges, 1995), but they examined various species belonging to brown (*Nereocystic luetkeana, Fucus fardneri, Costaria costata, Desmarestia viridis and Sargassum muticum*), green (*Ulva fenestrate and Codium fragile*) and red algae (*Opuntiella californica, Odonthalia floccose*). In their study, there were 11 kinds of aromatic products after CuO oxidation, and m-hydroxybenzoic acid and 3,5-dihydroxybenzoic acid were determined as non-amino acid derived materials as described above. The contents of these two products in body weight of algae were the highest in the brown algae (1.5-2.3 and 4.3-14 times than red and green algae for m-hydroxybenzoic acid and 3.5-dihydroxybenzoic acid), suggesting that production of phenolic compounds by brown algae is larger than other macroalgal groups.

4. Carbohydrates

4.1 Release of carbohydrates

Macroalgal body is covered by sticky mucus due to excretion of materials containing mucopolysaccharides for protection of their body from external stress such as desiccation (Percival & McDowell, 1981; Painter, 1983) and changes in ambient ion condition (Kloareg & Quatrano, 1988). Since the monosaccharide composition of mucopolysaccharides is different among brown, red and green algae, monosaccharide analysis is effective to understand the original source of mucopolysaccharide. Here we described the chemical characteristics of carbohydrate species originated from these algae, and reviewed the methodological aspects of monosaccharide analysis.

4.1.1 Brown algae

The major mucopolysaccharide of brown algae is alginate, which is made up of two kinds of uronic acids (mannuronic acid and guluronic acid). The molar ratio of mannuronic acid to guluronic acid (M/G) ranges from 0.25 to 2.25 (Kloareg & Quatrano, 1988). Another well-known mucopolysaccharide is fucan, which comprises L-fucose and sulphate as major constituents. Actually, fucans are heterogeneous group of polysaccharide, and the fucose content ranges from 55 to 96% of total monosaccharides (Marais & Joseleau, 2001; Bilan et al., 2004; 2006). Considering these monosaccharide composition, quantification of uronic acids and fucose would be available for characterization of mucopolysaccharides of brown algae.

4.1.2 Red algae

Carrageenans are mainly extracted from *Chondrus, Gigartina, Eucheuma* and *Hypnea*, and they are highly sulphated galactans (De Ruiter & Rudolph, 1997). These polysaccharides are categorized into several families based on the position of the sulphate groups. Agars are also a kind of galactan, but this is a low sulphated polymer, which is often extracted from *Gelidium, Gracilaria, Ahnfeltia, Acanthopeltis* and *Pterocladia* (Kloareg & Quatrano, 1988). These polysaccharides are commonly found in red algae, and galactose is the major constituents.

4.1.3 Green algae

Mucopolysaccharides originated from green algae are highly branched sulphated heteropolysaccharides such as xylogalactoarabinans, glucuronoxylorhamnans and rhamnoxylogalacto-galacturonan (Kloareg & Quatrano, 1988).

4.2 Depolymerization and derivatization of carbohydrates

4.2.1 Depolymerization

As mentioned above, monosaccharide composition reflects the original sources of the mucopolysaccharides released into extracellular region. For GC analysis of monosaccharide composition, it is necessary to depolymerize the polysaccharides and derivatize the monosaccharides in volatile forms. Well-known depolymerization and derivatization procedures are 1) hydrolysis-alditol-acetate or -trimethylsilyl and 2) methanolysis-trimethylsilyl methods. In this section, we describe the features of the methods, and introduce the application to macroalgal carbohydrates.

4.2.1.1 Acid hydrolysis

Various acid solutions have been examined for hydrolytic depolymerization of polysaccharides in marine samples, and detail of the results was previously reviewed (Panagiotopoulos & Sempere, 2005). Briefly, acid solutions which are mostly examined are HCl and H_2SO_4, and the recovery has been evaluated. Mopper (1973) compared these two acid solutions at same concentration (2 N), and suggested that HCl efficiently depolymerizes carbohydrate in ancient sediments but it led destruction of those in anoxic sediment (Mopper 1977). Acid strength is also important factor controlling the recovery yields. Two step of hydrolysis reaction, in which pre-treatment was carried out in 72% H_2SO_4 solution at ambient temperature and diluted sample (1-2 N) was heated at 100°C, are often used in order to achieve complete hydrolysis. These two steps reactions would be relatively strong, and sometimes induce loss of recovery of pentoses (Mopper, 1977), but the total yields of aldoses tend to be higher (Skoog & Benner, 1997). Mild hydrolysis was performed in solutions of dilute H_2SO_4, HCl, $CHCl_2COOH$, H_3PO_4, $(COOH)_2$ and trifluoacetic acid. In some cases, their recoveries are comparable with those of strong hydrolysis reaction (Panagiotopoulos & Sempere, 2005). However, they would be inappropriate for refractory species of carbohydrates such as cellulose (Skoog & Benner, 1997).

4.2.1.2 Derivatization and GC analysis

Since carbohydrates are polyhydroxy compounds, it is essential to convert them into the volatile derivatives. Commonly used derivatization methods are trimethylsilylation and alditol-acetate derivatizations. Generally, hexamethyldisilazane and trimethylchlorosilan have been used as derivatizating agents and pyridine as solvent for trimethylsilylation. Analytical procedures of this technique are simple and rapid (Sweeley et al., 1963), and appropriate volatility is obtained. This technique would be also applicable to nonreducing sugars. However, monosaccharide with free carbonyl groups can be present as different tautomers, and each tautomeric form occurs as different peak. Consequently, such derivative method would provide complicated chromatogram in GC analysis. When complex sugar mixture is analyzed, reduction of the monosaccharides should be considered to avoid the overlapping of peaks (Ruiz-Matute et al., 2011). In alditol-acetate method, the carbonyl group of monosaccharide is reduced using reducing reagent (e.g., KBH_4), and hydroxyl group of generated alditol is acetylated. This technique has several advantages, that alditol-acetate derivative produces single peak for each monosaccharide, and that the derivative is stable allowing clean-up for analysis (Knapp 1979). However, this method needs the large number of steps in the experimental procedures, and it is laborious and time consuming (Ruiz-Matute et al., 2011).

4.2.1.3 Methanolysis

Although hydrolysis is commonly used for analysis of neutral aldoses, the method is inappropriate for some kinds of carbohydrates such as uronic acid because of instability of uronic acid in acid hydrolysis reaction (Blake and Richards, 1968). Considering that uronic acid is also an important component of macroalgal mucopolysaccharides as mentioned above, alternative methodology should be examined. To overcome this problem, methanolysis reaction, which provides high recovery yields for uronic acid at 95-100% (Chambers and Clamp, 1971), is available.

Condition of methanolysis reaction commonly accepted is in 0.4-2 N methanolic HCl at 80°C for 5-24 h (Chambers & Clamp, 1971; Doco et al., 2001; Mejanelle et al., 2002). Under this condition, the recoveries of both neutral aldoses and uronic acid are stable (Chambers and Clamp, 1971). Since methanolysis reaction is interfered by water, this reaction should be carried out after drying the samples completely. After depolymerization of polysaccharides by methanolysis reaction, trimethylsilyl derivatization has been usually applied (Dierckxsens et al., 1983; Bleton et al., 1996; Doco et al., 2001; Mejanelle et al., 2002). Since water in the sample interferes in the trimethylsilyl reaction as well as methanolysis reaction, drying of the sample is an essential procedure (Chamber & Clamp, 1971)..

4.2.1.4 GC analysis for methanolysis-trimethylsilyl derivatives

In the methanolysis-trimethylsilyl method, several kinds of isomers are generated from one monosaccharide, and the chromatogram is often complicated due to the presence of a large number of peaks. Quantification of all of the isomers would be ideal, but detection of minor isomer peaks is sometimes difficult. In such a case, quantification is generally achieved by picking up major peaks, because the isomer composition would be constant regardless of initial chemical form of the monosaccharides if the methanolysis reaction was performed under same conditions (Mejanelle et al., 2002). Wada et al. (submitted) actually showed similar isomeric composition using seawater samples from natural environment and authentic standards.

The detail of GC or GCMS detection of methanolysis products has been already reviewed elsewhere (Mejanelle et al., 2004), and here we have simple explanation. When the analysis is carried out for less than 10 neutral and acidic monosaccharides, GC analysis will provide reliable identification of monosaccharide components. However, we should consider the presence of contaminant which is present in natural environment, because ambient seawater contains not only macroalgal exudates but organic constituents existing in seawater. In case that the contaminant peaks overlap with those of target monosaccharides, selective ion monitoring mode of GCMS instrument will be available (Wada et al., submitted). Using electron impact mode, some fragment ions are generated, and base peaks were m/z 73, 204 and 217 in the most cases. Considering that the peaks at m/z 204 and 217 are often found for pyranosides and furanosides, respectively (Mejanelle et al., 2004), the composition of fragment ions will provide useful information about not only quantification, but also identification of the isomers generated by methanolysis reaction.

4.3 Analysis of carbohydrates released from macroalgae

Mucopolysaccharides which are known as extracellular carbohydrates have been analyzed using GC or GCMS after some purification treatments (Lee et al., 2004; Mandal et al., 2007; Karmakar et al., 2010; Rioux et al., 2010; Stephanie et al., 2010). On the other hand, direct analysis for carbohydrates released into ambient seawater is limited (Wada et al., 2007; 2008). In their studies, *in situ* field bag experiment was developed, and the carbohydrates released from a brown alga, *Ecklonia cava*, were obtained by SCUBA diving. By hydrolysis and alditol-acetate derivative method, it was shown that fucose is the major monosaccharide component (36-44% of total carbohydrates), and that values had no significant seasonal variation. Since fucose is the major constituent of fucan derived from brown algae, constant dissolution of extracellular mucopolysaccharides into ambient environments from brown algae was suggested.

5. Conclusion

Macroalgae releases various organic compounds extracellularly and their role and dynamics in aquatic environments partly depend on the chemical composition. Since VHOC has been measured by numerous researchers, the VHOC analysis has been well improved in a few decades. On the other hand, there are possibilities of other unapplied option for phenolic compounds and carbohydrates analysis. For phenolic compounds, pyrolysis-GCMS is also considered as another potential tool, and Van Heemst et al. (1996) had tried to show the macroalgal contribution to marine organic matter pool. However, contamination of proteinaceous phenol under the process of pre-treatment was not as easy (Van Heemst et al., 1999). If this issue can be resolved in future, this instrument would also become a powerful tool for analysis of macroalgal phenolic compounds. Analysis of carbohydrates has been carried out for the extracts from algal body, but there is limited information on direct analysis of carbohydrates released to seawater as described above. Particularly, analysis using methanolysis method for seawater sample has been examined, yet. Thus, examination on the applicability of methanolysis method for seawater sample may be an important issue in the future.

6. Acknowledgement

This report is contribution no. 755 from the Shimoda Marine Research Center, University of Tsukuba.

7. References

Abdullah, M. I. & Fredriksen, S. (2004). Production, respiration and exudation of dissolved organic matter by the kelp *Laminaria hyperborea* along the west coast of Norway. Journal of Marine Biological Association of United Kingdom, Vol. 84, No. 5, pp. 887-894, ISSN 1469-7769

Abrahamsson, K. & Klick, S. (1990). Determination of biogenic and anthropogenic volatile halocarbons in sea water by liquid-liquid extraction and capillary gas chromatography. Journal of Chromatography, Vol. 513, pp. 39-45, ISSN 0021-9673

Abrahamsson, K., Ekdahl, A., Collén, J. & Pedersén M. (1995). Marine algae – a source of trichloroethylene and perchloroethylene. Limnology and Oceanography, Vol. 40, No. 7, (November 1995), pp. 1321-1326, ISSN 0024-3590

Abrahamsson, K. & Ekdahl, A. (1996). Volatile halogenated compounds and chlorophenols in the Skagerrak. Journal of Sea Research, Vol. 35, No. 1-3, (February, 1996), pp. 73-79, ISSN 1385-1101

Abrahamsson, K., Choo, K-S., Pedersén, M., Johansson, G. & Snoeijs, P. (2003). Effects of temperature on the production of hydrogen peroxide and volatile halocarbons by brackish-water algae. Phytochemistry, Vol. 64, No. 3, (October 2003), pp. 725-734, ISSN 0031-9422

Anderson, J. G., Toohey, D. W. & Brune, W. H. (1991). Free radicals within the Antarctic vortex: the role of CFCs in Antarctic ozone loss. Science, Vol. 251, No. 4989, (January 1991), pp. 39-46, ISSN 0036-8075

Baker, J. M., Sturges, W. T., Sugier, J. Sunnenberg, G., Lovett, A. A., Reeves, C. E., Nightingale, P. D. & Penkett, S. A. (2001). Emissions of CH_3Br, organochlorines, and organoiodines from temperate macroalgae. Chemosphere – Global Change Science, Vol. 3, No. 1, (January 2001), pp. 93-106, ISSN 1465-9972

Bengtsson, M. M., Sjøtun, K., Storesund, J. E. & Øvreås, L. (2011). Utilization of kelp-derived carbon sources by kelp surface-associated bacteria. Aquatic Microbial Ecology, Vol. 62, No. 2, (January 2011), pp. 191-199, ISSN 0948-3055

Bilan, M. I., Grachev, A. A., Ustuzhanina, N. E., Shashkov, A. S., Nifantiev, N. E. & Usov, A. I. (2004). A highly regular fraction of a fucoidan from the brown seaweed *Fucus distichus* L. Carbohydrate Research, Vol. 339, No. 3, (February 2004), pp. 511-517, ISSN 0008-6215

Bilan, M. I., Grachev, A. A., Shashkov, A. S., Nifantiev, N. E. & Usov, A. I. (2006). Structure of a fucoidan from the brown seaweed *Fucus serratus* L. Carbohydrate Research, Vol. 341, No. 2, (February 2006), pp. 238-245, ISSN 0008-6215

Blake, J. D. & Richards, G. N. (1968). Problems of lactonisation in the analysis of uronic acids. Carbohydrate Research, Vol. 8, No. 3, (November 1968), pp. 275-281, ISSN 0008-6215

Bleton, J., Mejanelle, P., Sansoulet, J., Goursand, S. & Tchapla, A. (1996). Characterization of neutral sugars and uronic acids after methanolysis and trimethylsilylation for recognition of plant gums. Journal of Chromatography A, Vol. 720, No. 1-2, (January 1996), pp. 27-49, ISSN 0021-9673

Blough, N. V. & Del Vecchio, R. (2002). Chromophoric DOM in the coastal environment, In: *Biogeochemistry of marine dissolved organic matter*, Hansell, D. A. & Carlson, C. A., (Eds.), 509-546, Academic Press, ISBN 0-12-323841-2, San Diego, USA

Bravo-Linares, C. M., Mudge, S. M. & Loyola-Sepulveda, R. H. (2010). Production of volatile organic compounds (VOCs) by temperate macroalgae. The use of solid phase microextraction (SPME) coupled to GC-MS as method of analysis. Journal of Chilean Chemical Society, Vol. 55, No. 2, (June 2010), pp. 227-232, ISSN 0717-9707

Chambers, R. E. & Clamp, J. R. (1971). An assessment of methanolysis and other factors used in the analysis of carbohydrate-containing materials. Biochemistry Journal, Vol. 125, No. 4, (December 1971), pp. 1009-1018, ISSN 0264-6021

Charpy-Roubaud, C. & Sournia, A. (1990). The comparative estimation of phytoplanktonic, microphytobenthic and macrophytobenthic primary production in the oceans. Marine Microbial Food Webs, Vol. 4, No. 1, pp. 31-57, ISSN 0297-8148

Crutzen, P. J. & Arnold, F. (1986). Nitric acid cloud formation in the cold Antarctic stratosphere: a major cause for the springtime 'ozone hole'. Nature, Vol. 324, No. 6098, (December 1986), pp. 651-655, ISSN 0028-0836

De Ruiter, G. A. & Rudolph, B. (1997). Carrageenan biotechnology, Trends in Food Science & Technology, Vol. 8, No. 12, (December 1997), pp. 389-395, ISSN 0924-2244

De Vooys, C. G. N. (1979).Primary production in aquatic environments. In: *The Global Carbon Cycle*, Bolin, B., Degens, E. T., Kempe, S. & Ketner, P., (Eds.), 259-292, Wiley, ISBN 978-0471997108, New York, USA

Dierckxsens, G. C., De Meyer, L. & Tonino, G. J. (1983). Simultaneous determination of uronic acids, hexosamines, and galactose of glycosaminoglycans by gas-liquid chromatography. Analytical Biochemistry, Vol. 130, No. 1, (April 1983), pp. 120-127, ISSN 0003-2697

Doco, T., O'Neill, M. A. & Pellerin, P. (2001). Determination of the neutral and acidic glycosyl-residue compositions of plant polysaccharides by GC-EI-MS analysis of the trimethylsilyl methyl glycoside derivatives. Carbohydrate Polymers, Vol. 46, No. 3, (November 2001), pp. 249-259, ISSN 0144-8617

Ekdahl, A. & Abrahamsson, K. (1997). A simple and sensitive method for the determination of volatile halogenated organic compounds in sea water in the amonl l^{-1} to pmol l^{-1} range. Analytica Chimica Acta, Vol. 357, No. 3, (December 1997), pp. 197-209 ISSN 0003-2670

Farman, J. C., Gardiner, B. G. & Shanklin, J. D. (1985). Large losses of total ozone in Antarctica reveal seasonal ClO_x/NO_x interaction. Nature, Vol. 315, No. 6016, (May 1985), pp. 207-210, ISSN 0028-0836

Giese, B., Laturnus, F., Adams, F. C. & Wiencke, C. (1999). Release of volatile iodinated C_1-C_4 hydrocarbons by marine macroalgae from various climate zones. Environmental Science and Technology, Vol. 33, No. 14, (June 1999), pp. 2432-2439, ISSN 0013-936X

Goñi, M. & Hedges J. I. (1995). Sources and reactivities of marine-derived organic matter in coastal sediments as determined by alkaline CuO oxidation. Geochimica et Cosmochimica Acta, Vol. 59, No. 14, (July 1995), pp. 2965-2981, ISSN 0016-7037

Goodwin, K. D., North, W. J. & Lidstrom, M. E. (1997). Production of bromoform and dibromomethane by Giant Kelp: Factors affecting release and comparison to anthropogenic bromine sources. Limnology and Oceanography, Vol. 42, No. 8, (December 1997), pp. 1725-1734, ISSN 0024-3590

Gschwend, P. M., MacFarlane, J. K. & Newman, K. A. (1985). Volatile halogenated organic compounds released to seawater from temperate marine macroalgae. Science, Vol. 227, No. 4690, (March, 1985), pp. 1033-1035, ISSN 0036-8075

Hatcher, B. G., Chapman, A. R. O. & Mann, K. H. (1977). An annual carbon budget for the kelp *Laminaria longicruris*. Marine Biology, Vol. 44, No. 1, (March 1977), pp. 85-96, ISSN 0025-3162

Jones, C. E., Hornsby, K. E., Dunk, R. M., Leigh, R. J. & Carpenter, L. J. (2009). Coastal measurements of short-lived reactive iodocarbons and bromocarbons at Roscoff, Brittany during the RHaMBLe campaign. Atmospheric Chemistry and Physics, Vol. 9, No. 4, pp. 8757-8769, ISSN 1680-7367

Karmakar, P., Pujol, C. A., Damonte, E. B., Ghosh, T. & Ray, B. (2010) Polysaccharides from *Padina tetrastromatica*: structural features, chemical modification and antiviral activity. Carbohydrate Polymers, Vol. 80, No. 2, (December 2009), pp. 513-520, ISSN 0144-8617

Khailov, K. M. & Burlakova, Z. P. (1969). Release of dissolved organic matter by marine seaweeds and distribution of their total organic production to inshore communities. Limnology and Oceanography, Vol. 14, No. 4, (July 1969), pp. 521-527, ISSN 0024-3590

Klick, S. (1992). Seasonal variations of biogenic and anthropogenic halocarbons in seawater from a coastal site. Limnology and Oceanography, Vol. 37, No. 7, (November 1992), pp. 1579-1585, ISSN 0024-3590

Kloareg, B. & Quatrano, R. S. (1988). Structure of the cell walls of marine algae and ecophysiological functions of the matrix polysaccharides. Oceanography and Marine Biology Annual Review, Vol. 26, pp. 259-315, ISSN 0078-3218

Knapp, D. R. (1979). Handbook of analytical derivatization reactions, Vol. 1, Wiley Interscience, ISBN 978-0-471-03469-8, New York, USA

Łabudzińska, A. & Gorczyńska, K. (1995). The UV difference spectra as a characteristic feature of phenols and aromatic amines. Journal of Molecular Structure, Vol. 349, (April 1995), pp. 469-472, ISSN 0022-2860

Laturnus, F. (1995). Release of volatile halogenated organic compounds by unialgal cultures of polar macroalgae. Chemosphere, Vol. 31, pp. 3387-3395.

Laturnus, F. (1996). Volatile halocarbons released from Arctic macroalgae. Marine Chemistry, Vol. 55, pp. 359-366.

Laturnus, F., Wiencke, C. & Klöser, H. (1996). Antarctic macroalgae-Sources of volatile halogenated organic compounds. Marine Environmental Research, Vol. 41, No. 2, pp. 169-181, ISSN 0141-1136

Laturnus, F., Wiencke, C. & Adams, F. C. (1998). Influence of light conditions on the release of volatile halocarbons by Antarctic macroalgae. Marine Environmental Research, Vol. 45, No. 3, (April 1998), pp. 285-294, ISSN 0141-1136

Laturnus, F., Giese, B., Wiencke, C. & Adams, F. C. (2000). Low-molecular-weight organoiodine and organobromine compounds released by polar macroalgae – The influence of abiotic factors. Fresenius Journal of Analytical Chemistry, Vol. 368, No. 2-3, (October 2000), pp. 297-302, ISSN 0937-0633

Laturnus, F. (2001). Marine macroalgae in polar regions as natural sources for volatile organohalogens. Environmental Science and Pollution Research International, Vol. 8, No. 2, pp. 103-108, ISSN 0944-1344

Laturnus, F., Svensson, T., Wiencke, C. & Öberg, G. (2004). Ultraviolet radiation affects emission of ozone-depleting substances by marine macroalgae: results from a laboratory incubation study. Environmental Science and Technology, Vol. 38, No. 24, (December 2004), pp. 6605-6609, ISSN 0013-936X

Laturnus, F., Svensson, T. & Wiencke, C. (2010). Release of reactive organic halogens by the brown macroalga *Saccharina latissima* after exposure to ultraviolet radiation. Polar Research, Vol. 29, No. 3, pp. 379-384, ISSN 1751-8369

Lee, J-B., Hayashi, K., Hashimoto, M., Nakano, T. & Hayashi, T. (2004). Novel antiviral fucoidan from sporophyll of *Undaria pinnatifida* (Mekabu). Chemical Pharmaceutical Bulletin, Vol. 52, No. 9, (September 2004), pp. 1091-1094, ISSN 0009-2363

Lovelock, J. E. (1975). Natural halocarbons in the air and in the sea. Nature, Vol. 256, No. 5514, (July 1975), pp. 193-194, ISSN 0028-0836

Mandal, P., Mateu, C. G., Chattopadhyay, K., Pujol, C. A., Damonte, E. B. & Ray, B. (2007). Structural features and antiviral activity of sulphated fucans from the brown seaweed *Cystoseira indica*. Antiviral Chemistry & Chemotherapy, Vol. 18, No. 3, pp. 153-162, ISSN 0956-3202

Manley, S. L. & Dastoor, M. N. (1987). Methyl halide (CH_3X) production from the giant kelp, *Macrocystis*, and estimates of global CH_3X production by kelp. Limnology and Oceanography, Vol. 32, No. 3, pp. 709-715, ISSN 0024-3590

Manley, S. L. & Dastoor, M. N. (1988). Methyl iodide (CH_3I) production by kelp and associated microbes. Marine Biology, Vol. 98, No. 4, pp. 477-482, ISSN 0025-3162

Manley, S. L., Goodwin, K. & North, W. J. (1992). Laboratory production of bromoform, methylene bromide, and methyl iodide by macroalgae and distribution in nearshore southern California waters. Limnology and Oceanography, Vol. 37, No. 8, (December 1992), pp. 1652-1659, ISSN 0024-3590

Manley, S. L. & Barbero, P. E. (2001). Physiological constraints on bromoform ($CHBr_3$) production by *Ulva lactuca* (Chlorophyta). Limnology and Oceanography, Vol. 46, No. 6, (September 2001), pp. 1392-1399, ISSN 0024-3590

Mann, K. H. (1973). Seaweeds: their productivity and strategy for growth. Science, Vol. 182, No. 4116, pp. 975-981, ISSN 0036-8075

Marais, M-F. & Joseleau, J-P. (2001). A fucoidan fraction from *Ascophyllum nodosum*. Carbohydrate Research, Vol. 336, No. 2, (November 2001), pp. 155-159, ISSN 0008-6215

Marshall, R. A., Harper, D. B., McRoberts, W. C. & Dring, M. J. (1999). Volatile bromocarbons produced by *Falkenbergia* stages of *Asparagopsis* spp. (Rhodophyta). Limnology and Oceanography, Vol. 44, No. 5, (July 1999), pp. 1348-1352, ISSN 0024-3590

Marshall, R. A., Hamilton, J. T. G., Dring, M. J. & Harper, D. B. (2000). The red alga *Asparagopsis taxiformis/Falkenbergia hillebrandii* – a possible source of trichloroethylene and perchloroethylene? Limnology and Oceanography, Vol. 45, No. 2 (March 2000) pp. 516-519, ISSN 0024-3590

Mejanelle, P., Bleton, J., Tchapla, A. & Goursaud, S. (2002). Gas chromatography-mass spectrometric analysis of monosaccharides after methanolysis and trimethylsilylation. Potential for characterization of substances of vegetal origin: application to the study of museum objects. Journal of Chromatography Library, Vol. 66, pp. 845-902, ISSN 0301-4770

Mopper, K. (1973). Aspects of the biogeochemistry of carbohydrates in aquatic environments. Thesis, M. I. T. – Woods Hole Oceanographic Institution Joint Program in Oceanography.

Mopper, K. (1977). Sugars and uronic acids in sediment and water from the Black Sea and North Sea with emphasis on analytical techniques. Marine Chemistry, Vol. 5, No. 4-6, (November 1977), pp. 585-603, ISSN 0304-4203

Mtolera, M. S. P., Collén, J., Pedersén, M., Ekdahl, A., Abrahamsson, K. & Semesi, A. K. (1996). Stress-induced production of volatile halogenated organic compounds in *Eucheuma denticulatum* (Rhodophyta) caused by elevated pH and high light intensities. European Journal of Phycology, Vol. 31, No. 1, pp. 89-95, ISSN 0967-0262

Newman, K. A. & Gschwend, P. M. (1987). A method for quantitative determination of volatile organic compounds in marine macroalgae. Limnology and Oceanography, Vol. 32, No. 3, (May 1987), pp. 702-708, ISSN 0024-3590

Nightingale, P. D., Malin, G. & Liss, P. S. (1995). Production of chloroform and other low-molecular-weight halocarbons by some species of macroalgae. Limnology and Oceanography, Vol. 40, No. 4, (June 1995), pp. 680-689, ISSN 0024-3590

Painter, T. J. (1983). Algal polysaccharides. In: *The polysaccharides*, Aspinall, G. O., (Ed.), Vol. 2, 195-285, Academic Press, ISBN 0-12-065602-7, New York, USA

Panagiotopoulos, C. & Sempéré, R. (2005). Analytical methods for the determination of sugars in marine samples: a historical perspective and future directions. Limnology and Oceanography: Methods, Vol. 3, (January 2005), pp. 419-454, ISSN 1541-5856

Paul, V. J., Puglisi, M. P. & Ritson-Williams, R. (2006). Marine chemical ecology, Natural Product Reports, Vol. 23, (February 2006), pp. 153-180, ISSN 0265-0568

Pedersén, M., Collén, J., Abrahamsson, K. & Ekdahl, A. (1996). Production of halocarbons from seaweeds: an oxidative stress reaction? Scientia Marina, Vol. 60, pp. 257-263, ISSN 0214-8358

Percival, E. & McDowell, R. H. (1981). Algal cell walls-composition and biosynthesis. In: Encyclopedia of plant physiology, Tanner, W. & Loewus, F. A., (Eds.), Vol. 13, 277-316, Springer, ISBN 0-387-11007-0, New York, USA

Rioux, L-E., Turgeon, S. L. & Beaulieu, M. (2010). Structural characterization of laminaran and galactofucan extracted from the brown seaweed *Saccharina longicruris*. Phytochemistry, Vol. 71, No. 13, (September 2010), pp. 1586-1595, ISSN 0031-9422

Ruiz-Matute, A. I., Hernández-Hernández, O., Rodríguez-Sánchez, S., Sanz, M. L. & Martínez-Castro, I. (2011). Derivatization of carbohydrates for GC and GC-MS analyses. Journal of Chromatography B, Vol. 879, No. 17-18, (May 2011), pp. 1226-1240, ISSN 1570-0232

Schall, C., Laturnus, F. & Heumann, K. G. (1994). Biogenic volatile organoiodine and organobromine compounds released from polar macroalgae. Chemosphere, Vol. 28, No. 7, (April 1994), pp. 1315-1324, ISSN 0045-6535

Shibata, T., Hama, Y., Miyasaki, T., Ito, M. & Nakamura, T. (2006). Extracellular secretion of phenolic substances from living brown algae. Journal of Applied Phycology, Vol. 18, No. 6, (December 2006), pp. 787-794, ISSN 0921-8971

Sieburth, J. M. (1969). Studies on algal substances in the sea. III. The production of extracellular organic matter by littoral marine algae. Journal of Experimental Marine Biology and Ecology, Vol. 3, No. 3, pp. 290-309, ISSN 0022-0981

Skoog, A. & Benner, R. (1997). Aldoses in various size fractions of marine organic matter: implications for carbon cycling. Limnology and Oceanography, Vol. 42, No. 8, (December 1997), pp. 1803-1813, ISSN 0024-3590

Solomon, S. (1990). Progress towards a quantitative understanding of Antarctic ozone depletion. Nature, Vol. 347, No. 6291, (September 1900), pp. 347-354, ISSN 0028-0836

Stephanie, B., Eric, D., Sophie, F. M., Christian, B. & Yu, G. (2010) Carrageenan from *Solieria chordalis* (Gigartinales): structural analysis and immunological activities of the low molecular weight fractions. Carbohydrate Polymers, Vol. 81, No. 2, (June 2010), pp. 448-460, ISSN 0144-8617

Sweeley, C. C., Bentley, R., Makita, M & Wells, W. W. (1963). Gas-liquid chromatography of trimethylsilyl derivatives of sugars and related substances. Journal of the American Chemical Society, Vol. 85, No. 16, pp. 2497-2507, ISSN 0002-7863

Van Alstyne, K. L. (1988). Herbivore grazing increases polyphenolic defenses in the intertidal brown alga *Fucus distichus*. Ecology, Vol. 69, No. 3, 655-663, ISSN 0012-9658

Van Heemst, J. D. H., Peulvé, S. & De Leeuw, J. W. (1996). Novel algal polyphenolic biomacromolecules as significant contributors to resistant fractions of marine dissolved and particulate organic matter. Organic Geochemistry, Vol. 24, No. 6-7, (June 1996), pp. 629-640, ISSN 0146-6380

Van Heemst, J. D. H., Van Bergen, P. F., Stankiewicz, B. A. & De Leeuw, J. W. (1999). Multiple sources of alkylphenols produced upon pyrolysis of DOM, POM and recent sediments. Journal of Analytical and Applied Pyrolysis, Vol. 52, No. 2, (November 1999), pp. 239-256, ISSN 0165-2370

Wada, S., Aoki, M. N., Tsuchiya, Y., Sato, T., Shinagawa, H. & Hama, T. (2007). Quantitative and qualitative analyses of dissolved organic matter released from Ecklonia cava Kjellman, in Oura bay, Shimoda, Izu Peninsula, Japan. Journal of Experimental Marine Biology and Ecology, Vol. 349, No. 2, (October 2007), pp. 344-358, ISSN 0022-0981

Wada, S., Aoki, M. N., Mikami, A., Komatsu, T., Tsuchiya, Y., Sato, T., Shinagawa, H. & Hama T. (2008). Bioavailability of macroalgal dissolved organic matter in seawater. Marine Ecology-Progress Series, Vol. 370, (October 2008), pp. 33-44, ISSN 0171-8630

Weinberger, F., Coquempot, B., Forner, S., Morin, P., Kloareg, B. & Potin, P. (2007). Different regulation of haloperoxidation during agar oligosaccharide-activated defense mechanisms in two related red algae, *Gracilaria* sp. and *Gracilaria chilensis*. Journal of Experimental Botany, Vol. 58, No. 15-16, pp. 4365-4372, ISSN 0022-0957

Yokohama, Y., Tanaka, J. & Chihara, M. (1987). Productivity of the *Ecklonia cava* community in a bay of Izu Peninsula on the Pacific coast of Japan. The Botanical Magazine, Tokyo, Vol. 100, No. 2pp. 129-141, ISSN 0918-9440

Part 2

Selected Applications of Gas Chromatography in Industrial Applications

5

Application of Gas Chromatography in Monitoring of Organic and Decontamination Reactions

Pranav Kumar Gutch

Defence Research & Development, Establishment Jhansi Road, Gwalior,
India

1. Introduction

Gas chromatography (GC) is a separation technique commonly used for organic compounds which can be converted into gaseous phase without decomposition. The components to be separated are carried through the column by an inert gas (Carrier gas). The separation of the analyte usually takes place in a long capillary tube (the column). The wall of the column is coated with a thin film of a very high boiling liquid (stationary phase), through which an inert gas (mobile phase) flows. The component of a mixture are portioning between the two phases. The stronger the interaction of a component with the stationary phase, the longer will be the retention. The rate of movement of the various components along the column depends upon their tendency to absorb in the stationary phase. Component of mixture are separated at different time interval and it give only a single peak for each separated component of mixture. A plot of the time necessary for a compound to move through the column (retention time) versus its amount (intensity) is called a chromatogram. The time for injection point to peak maxima is called retention time. Retention time depends on various factor viz Flow rate of carrier gas, Temp of column, and nature of stationary phase.

2. Monitoring of organic reactions by GC

GC (Pattision 1973) is generally used for determination of purity, identification of samples and monitoring of organic reactions. Decontamination reaction of chemical warfare agents can also be monitored by GC. N-Chloro compounds (Kovari, E. 2007) have been extensively exploited, both for fundamental research and a wide range of industrial applications, (Koval, I. 2002, 2001) owing to their easy handling, commercial availability and high storage stability. As a result, intensive research and studies have been carried out over long period of time on their chlorination, (Ghorbani, Vaghei. 2009) oxidation, water disinfection (Bogoczek, R. 1989)& some other applications (Kowalski, P. 2005) in synthetic organic chemistry. N-chloro compounds (Singh, R. 2011) have also been reported for decontamination of chemical warfare agents.

The polymeric N-Chloro resin proved to be very strong chloridine agent and found to be easily reusable. The additional advantage of polymeric chloramines (Gutch, P. 2011) in higher stability as compared to low chloramines such as chloramines-T and dichloramine – T. Active chlorine content does not decrease with prolonged storage time. Hetrogenised reagent has evinced the attention of researchers because of their handling ease, lower toxicity, nonexplosive, and malodorous nature. Another impotent motive of supported reagent is facile isolation of pure product than their solutions phase equivalents.

N-chloramines (Gutch, P. 2011) in which chlorine is directly attached to nitrogen can generate positively charge chlorine (Cl^+) which is an oxidising species. N-Chloro compound macroporous poly(styrene-co-divinylbenzene) resin having N,N-dichlorosulfonamide groups have been used as a polymer supported reagent for chlorination, oxidation for residual sulfides, cyanides, thiocyanates, water disinfection and some application in synthetic organic chemistry. In recent year the use of recyclable reagents has received considerable interest in organic synthesis. This prompted us to explore the possibility of using the stable, non toxic, recyclable, and efficient positive chlorine releasing reagent, N, N-dichloro poly (styrene-co-divinylbenzene) sulfonamide polymeric beads for oxidation, chlorination and decontamination reactions.

2.1 Synthesis of dialkylchlorophosphate from dialkylphosphites at room temperature

Organophosphorus compounds (OPs) (Eto, M. 1974) have attracted the attention of researchers because of their wide applications in industrial, agricultural, biochemical, and medicinal areas. Naturally occurring OPs play an important role in the maintenance of life processes. It is interesting to note that chemical, physical, and biological properties of OPs are governed by the stereochemical disposition of substituents around the phosphorous atom. One such class of compounds is the dialkylchlorophosphates. These chlorophosphates are used for the transformation of various functional groups. Recently, diethylchlorophosphate has been used as an efficient reagent in cyclization reactions and in regioselective ring opening of epoxides.

Methods described in the literature for the preparation of dialkylchlorophosphates, involve chlorination of the corresponding phosphites (dialkylphosphites/trialkylphosphites) with reagents such as elemental chlorine, phosgene, SO_2Cl_2, S_2Cl_2, SCl_2, CCl_4, CCl_3NO_2, $PhSO_2NCl_2$, C_2Cl_6, $ClSCCl_3$, $CuCl_2$, perchlorofulvalene, and N-chlorosuccinimide. Among these procedures, only a few can be considered as convenient laboratory methods for the synthesis of dialkylchlorophosphates. Most of these reported methods either use reagents or produce undesired by-products, which are difficult to remove from the sensitive chlorophosphates, while others are time consuming, involve expensive and unstable reagents, and require harsh conditions.

Having established a commercially synthetic procedure for N, N dichloro poly (styrene-co-divinylbenzene)(Gutch, P. 2007) sulfonamide polymeric beads(we investigated its use as an alternative reagent for the conversion of di alkylphosphite to dialkylchlorophosphate (Gupta, Hemendra. 2008).This development enable us to obtained almost all quantitative yields of products in short reaction times. The room temperature (20-25°C) reaction of various dialkylphosphite with N, N dichloro poly (styrene-co-divinylbenzene) sulfonamide polymeric beads afforded the corresponding dialkylchlorophosphates in 3-4 h in excellent yields (Scheme 1).The reaction was monitored by removing an aliquot and analyzing by GC-MS. The progress of a typical reaction as monitored by GC-MS is depicted in fig. 1.

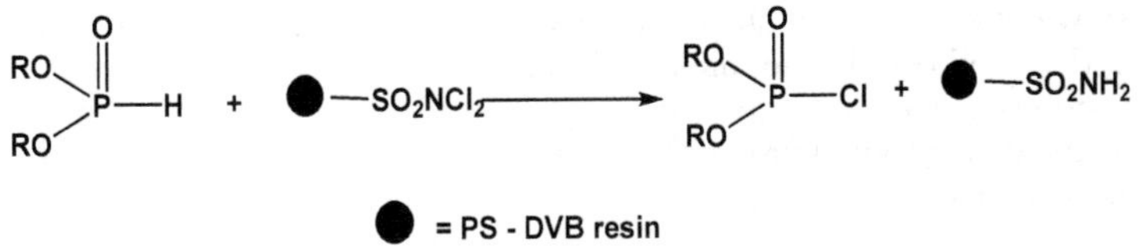

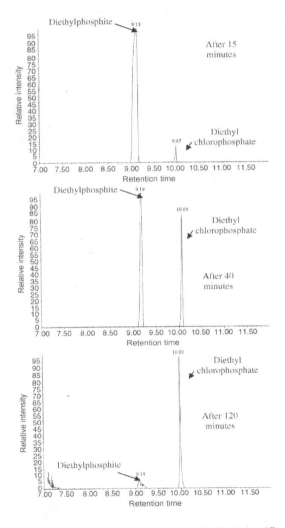

Scheme 1. Dialkylchlorophosphateprepared from dialkylphosphites using polystyrene divinylbenzene (PS-DVB) bound reagent.

Figure 1. Progress of a typical reaction as monitored by GC–MS analysis at different time intervals.

Fig. 1. Progress of a typical reaction as monitored by GC-MS analysis at different time intervals.

In conclusion, we have described an efficient reagent for the rapid and convenient conversion of dialkylphosphites to dialkylchlorophosphates under mild conditions using N,N-dichloro poly(styrene-co-divinyl benzene) sulfonamide polymeric beads, as a stable and non-toxic reagent at room temperature. Here in this work it is proved to be an important and valuable reagent.

2.2 Synthesis of dicyanooxiranes derived from benzylidenemalononitriles

BenzylideneMallononitriles(Rose, S. 1969) have received much attention as cytotoxic agents against tumors and some derivatives have also been used as rodent repellents. 2-Chlorobenzylidene Malononitrile (CS) (Jones, G. 1972) is one of the most potent lachrymator and skin irritants used as riot control agent. Benzylidinemalononitriles (Gutch, PK, Thesis 1997) are highly sensitive to oxygen and form Gem-dicyano epoxides which constitute an important class of synthetic intermediates. These epoxide react with a variety of a reagent such as alcohol, water, hydrazine, and hydroxylamine in presence of halogen acid to give useful products including hydroxamic acid which is an interesting material for the synthesis of intermidiatesaziridinones used in situ as precursor to α- hydroxy and α –amino acids.We have developed an improved methods for the synthesis of substituted phenyl-1,1dicyanooxiranes (Scheme 2) from benzylidenemalononitrile using calcium hypochlorite (Gutch, P. March 2001) in carbon tetrachloride at 25°C. The reaction completion within 90 min and results in 98-99% yields of dicyanooxiranes. The reaction does not require any maintenance of the pH and no side products are formed. The reaction was monitored by GC (fig. 2) using FID as a detector at 140 °C

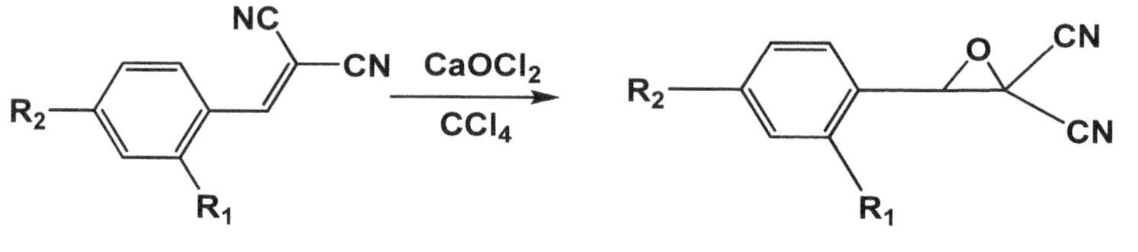

	R$_1$	R$_2$
1.	H	H
2.	Cl	H
3.	Br	H
4.	CH$_3$	H
5.	NO$_2$	H
6.	H	Cl
7.	H	Br
8.	H	CH$_3$
9.	H	NO$_2$

Scheme 2. Synthesis of substituted phenyl-1, 1'-dicyanooxiranes

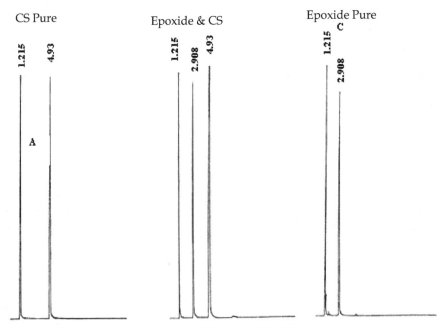

Fig. 2. Reaction of CS with calcium hypochlorite on DB-1701 column at 140 °C: Chromatogram: A= Reaction mixture at zero time showing only CS: B=Reaction Mixture after 20 minutes showing the formation Epoxide: C=Reaction mixture after 45 minutes showing disappearance of CS.

2.3 Synthesis of disulfide from thiols

Disulfides (Pathak, U. 2009)) have found industrial applications as vulcanizing agents and are important synthetic intermediates with many applications in organic synthesis. Disulfides are primarily produced from thiols, which are readily available and facile to prepare. Oxidative coupling of thiols to disulfides under neutral and mild conditions is of practical importance in synthetic chemistry and biochemistry. However, synthesis of disulfides from thiols sometimes can be problematic due to the over oxidation of thiols to sulfoxides and sulfones. On the other hand, many successful reagents such as 1,3-dibromo-5,5-dimethylhydantoin, cerium(IV) salts, permanganates, transition metal oxides, sodium chlorite, peroxides, halogens, solid acid reagents, monochloro poly(styrenehydantoin)(Akin, Akdag. 2006) beads, N-Phenyltriazolinedione, poly(N-bromomaleimide), (Bahman, Tamani. 2009) have been developed for the synthesis of disulfides from thiols, under a range of experimental conditions.

A novel method for oxidative coupling of thiols (Gutch, P. 2011) to their corresponding disulfides (scheme 3) using N, N-dichloro poly (styrene-co-divinyl benzene) sulphonamide (PS-DVS) beads was developed.In this study, we have employed N, N- dichloro poly (styrene-co-divinylbenzene) sulfonamide beads in water for the conversion of thiols to symmetric disulfides. These polymeric beads did not over oxidize the thiols to sulfoxides or sulfones and the sole product isolated from these experiments was disulfides. Presumably, the reactions proceed through chlorine transfer from amide N-halamine to thiols to yield disulfides and hydrochloric acid.

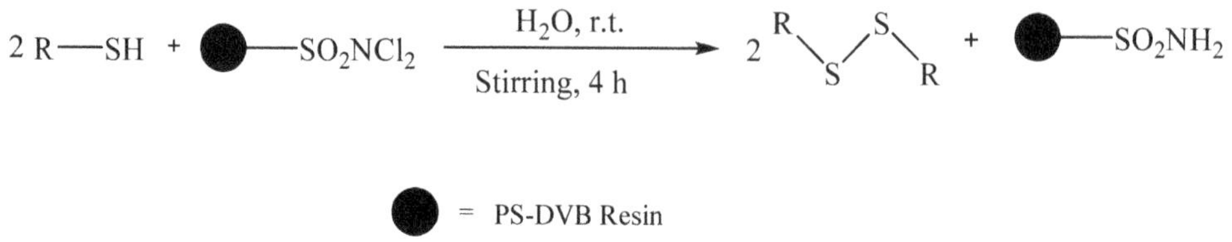

$$2\ R\!-\!SH\ +\ \bullet\!-\!SO_2NCl_2\ \xrightarrow[\text{Stirring, 4 h}]{\text{H}_2\text{O, r.t.}}\ 2\ \underset{S}{\overset{R}{\diagdown}}\!\underset{R}{\overset{S}{\diagup}}\ +\ \bullet\!-\!SO_2NH_2$$

$$\bullet\ =\ \text{PS-DVB Resin}$$

Scheme 3. Synthesis of disulfides from thiols

Thiol(0.05 mol) was added to a suspension of N, N-dichloro poly (styrene-co-divinyl benzene) sulphonamide beads (containing 0.05 mol of active chlorine) in water (30mL). The mixture was stirred at room temperature for 4 hrs and reaction was monitored by GC (Fig. 3). PS-DVS beads act as novel and selective oxidative agents which efficiently reduce the reaction time, increase the product yield without producing over oxidized products and performed under air atmosphere.

In conclusion N, N-dichloro poly (styrene-co-divinylbenzene) sulfonamide resin is an efficient reagent for fulfill different required objective in organic synthesis. Here in this work it proved to be an important and valuable reagent for the corresponding disulfide. The conversion was very efficient, with excellent yield and polymeric reagent can be recovered activated and reused. The main advantages of the method is that the reaction was very clean and operationally simple. Therefore the method is very attractive for organic chemist.

3. Monitoring of decontamination reaction of chemical reaction for chemical warfare agent

Decontamination (Yu-chu, Yang. 1992) is an important unavoidable part in protection against Chemical warfare agent. The aim of decontamination is to rapidly and effectively render harmless or remove poisonous substances both on personal and equipment. Though a variety of decontaminating agents have been reported over the years, however most of them suffer from drawbacks such as the use of hazardous solvents and prolonged decontamination time. Further, these reagents cannot be used on skin due to the toxicity of solvent and reagents.

Decontamination (Somani, S. (Ed). 1992) may be defined as a method of conversion of toxic chemicals into harmless products which can be handled safely, either by degradation or detoxification using suitable reagents. Decontamination plays a vital role in defence against chemical warfare agents. The toxic materials must be eliminated from the battlefield etc. by application of some efficient decontamination methods as fast as possible for resuming routine activities. Decontamination (Talmage, S. 2007) of CW agents is required on the battlefield, in laboratories, pilot plants and places of chemical agent production, storage, and destruction sites.

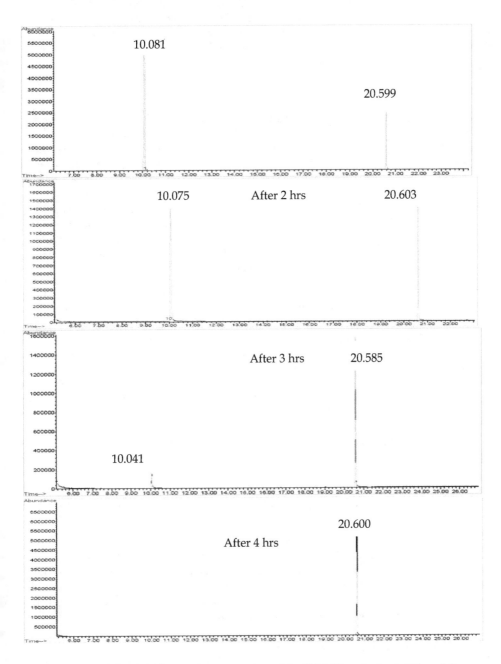

Fig. 3. Progress of disulphide reaction as monitored by GC-MS analysis at different time interval

Toxic substances can also be decontaminated (Yang, Y. 1995) by chemical modification of their toxic structures by using small compounds like hydrogen peroxide. Decontamination (Dubey, D. 1999) through chemical reactions is more reliable and effective as compared to physical methods. Chemical reaction yields less toxic or non-toxic products, while in the physical process they are just removed from the site.

3.1 Simulants

As with all chemical warfare agents, testing with the actual agents which are often very toxic is difficult to synthesize, their handling is also difficult, and they are expensive and are not commercially available. The developmental work on their process is often done on surrogate chemical that possesses many of the important features of the agent. Surrogate, also called a simulant or analog, must be carefully chosen to reasonably substitute for the target agent in the specific reaction pathway or reaction conditions under investigation. In the case of the partial oxidation of sulfur mustard, for example, an appropriate surrogate should clearly have a sulfide bond. Ideally, the surrogate would have similar structural features allowing for the same reaction mechanisms that would detoxify the agent, and would also have similar physical properties like solubility and vapor pressure. Also, one should consider secondary reactivity when possible, allowing for any foreseeable side reaction.

Simulants are those compounds which are less toxic but they have same chemical reactivity like toxicants. For the standardization of the chemical reactions, simulants can be used in place of actual agents. In order to gain a more complete understanding of CW agent chemistry, it is often necessary to study the reactions of a series of agent-analogs under the same conditions. For example, the monofunctional derivatives of mustard, $RSCH_2CH_2Cl$ (R= Me, Et, Ph), $RSCH_2CH_2X$ (X = tosylate, brosylate, Br-, I-, or other leaving groups), react via the same mechanism as those of HD, but their reaction products and kinetics rate expressions are much simpler. 2-chloroethylphenylsulfide, 2-chloroethyl ethylsulfide (half mustard) and dibutyl sulfide (DBS) are all simulants of HD. The use of simulants makes it easier to isolate the variable that affects the agent chemistry. VX analogs, (C_2H_5S) (CH_3) P (O)-(C_2H_5O), $(C_2H_5S)(C_6H_5)P(O)$-(C_2H_5O) have also been studied to isolate the effect of the diisopropylamino group on the reaction chemistry of VX. Simulants are those compounds which are less toxic but they have same chemical reactivity like HD & VX. For the standardization of the chemical reactions, simulants can be used in place of actual agents. 2-chloroethyl phenyl sulphide& methyl phosphothioic acid O, S- diethyl ester are simulants of HD & VX respectively.

Our aim in this work is to overcome the limitation and drawbacks of the reported decontaminants (Prasad, G. (2009)). It has been in corrupted as a reactive ingredient in a formulation developed in our laboratory to decontaminated bis (2-chloroethyl) sulfide (SM), a chemical warfare agent.N,N-Dichloro poly(styrene-codivinyl benzene) sulfonamide, a new class of readily available, economical, commercially viable, and recyclable chlorinating reagent was chosen for use as a decontaminating agent in this study.

3.2 Decontamination of Sulfur Mustard (SM)

We have already reported N, N-dichloro poly (styrene-co-divinyl benzene) sulphonamide polymeric beads for decontamination of simulant of SM. N,N-dichloro poly(styrene-co-divinylbenzene) sulphonamide reacts with 2-chloro ethyl phenyl sulphide, a simulant of sulfur mustard (SM), at room temperature, yielding corresponding nontoxic sulfone and sulfoxide in aqueous as well as aprotic medium.

Sulfur mustard (SM) is a cytotoxic and persistent blistering agent that was used as a weapon of mass destruction in the World War I and recently in the Iraq-Iran war (1980-1988). Detoxification (Gutch, P. 2008) of the chemical warfare agent (CWA), sulfur mustard is one of the primary goals to get rid of its toxic effects. An efficient and operationally simple method is developed for chemical decontamination of sulfur mustard. A new chlorine bearing reagent N,N-dichloro poly (styrene-co-divinyl benzene) sulfonamide was developed to deactivate the sulfur mustard (**Scheme 4**) in aqueous (acetonitrile: water) medium. This decontamination reaction was monitored by GC-FID (Fig.4) and the products were analyzed by GC-MS (Fig.5).

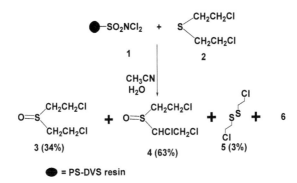

Scheme. 4. Reaction of SM with polymeric Beads in aqueous medium

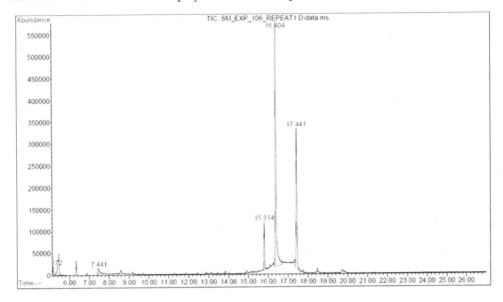

Fig. 4. Gas Chromatogram of SM after decontamination

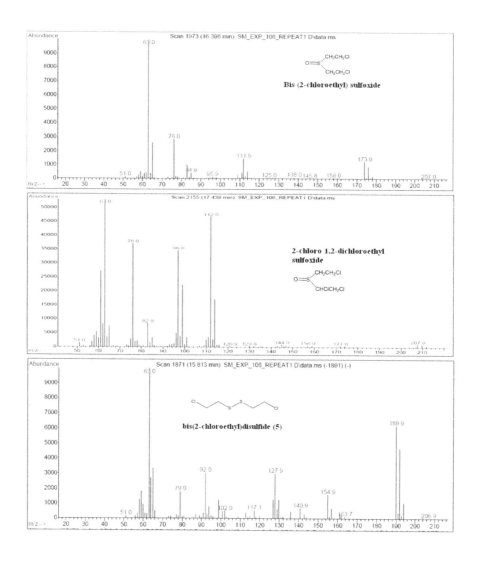

Fig. 5. Mass spectrum of decontaminated products (3-5) of SM in aqueous medium

N, N-dichloro poly (styrene-co-divinyl benzene) sulphonamide (Gutch, P. 2011) polymeric beads (3.0 mmol) was added slowly to a stirred solution of SM (1.0 equiv.) in acetonitrile water mixture (5 ml) at room temperature. It was found that 100 % SM was decontaminated instantaneously in aqueous medium and decontaminated products were identified by Gas Chromatography-Mass Spectrometry. The GC conditions used were as follows; column HP-

5 (30m× 0.250mm i.d., 0.25 μm film thickness) with a temperature programme of 50°C for 2 min followed by a linear gradient to 250°C at 10°Cmin⁻¹, and hold at 250°C for 5 min. The injector temperature was maintained at 250°C.

The decontamination reaction of SM is studied with varying molar concentration of active chlorine available in PS-DVB. At equimolar ratio of SM and active chlorine on PS-DVB, the major product is bis (2-chloroethyl) sulfoxide(3). At 1:2 molar ratio, in addition to (3), 2-chloro 1,2-dichloroethyl sulfoxide(4) is also formed; whereas, at 1:4 molar ratio the major product is 2-chloro 1, 2-dichloroethyl sulfoxide (4) with a trace amount of bis(2-chloroethyl)disulfide (5). In all the reactions N, N-dichloro poly (styrene-co-divinyl benzene) sulfonamide (PS-DVB) polymeric beads (1) were converted into poly (styrene-co-divinyl benzene) sulfonamide (PS-DVB) polymeric beads (6).

This reagent has advantage over earlier reported reagents in terms of effectiveness, stability, non toxicity, and cost, ease of synthesis, recyclability (collected after filtration, rechlorinated and used for further reaction) and instantaneous decontamination of sulfur mustard at room temperature.

3.3 Decontamination of S-2-(diisopropylamino) ethyl O-ethyl methylphosphonothioate(VX)

Organophosphorus compounds (OPs) have been widely used in the industry, veterinary and human medicine, in the agriculture as pesticides or can be misused for military purpose i.e. as Chemical Warfare Agents (CWA). The chemical warfare nerve agents, commonly known as nerve gases, are in fact not gases but polar organic liquids at ambient conditions. Most of these nerve agents are P (V) organophosphorus esters that are similar in structure to an insecticide.

Highly toxic organophosphorus compounds (OPCs) that are nerve agents such as sarin, soman, tabun and cyclosarin are lethal chemical warfare agents, also known as G-agents and were developed by G. Shrader in Germany during World War-II.It was then followed by the discovery of VX (S-2-(diisopropylamino) ethyl O-ethyl methylphosphonothioate), a nerve agent that was much more potent than all the known G-agents. Currently there are two known V-agent stockpiles; VX (S-2-(diisopropylamino) ethyl O-ethyl methylphosphonothioate), with thousands of tons in the USA, and an analogue and isomer, RVX (Russian-VX, S-2-(diethylamino) ethyl-O-isobutyl methylphosphonothioate) in Russia.

We investigated its use as an alternative reagent for decontamination of simulant of chemical warfare agent VX at room temperature. In order to gain more complete understanding of agent chemistry, a simulant of VX; O, S-diethyl methylphosphonothiolate (OSDEMP) was synthesized initially for decontamination study.

An efficient and operationally simple method is developed for chemical decontamination of simulant of VX. N, N-dichloro poly (styrene-co-divinyl benzene) sulfonamide was developed to deactivate the simulant of VX (Scheme 5) in aqueous medium. This decontamination reaction was monitored by gas chromatography.

The decontamination of CWAs from structure, environment, media, and even personal has become an area of particular interest in recent years because of increased homeland security concern. In addition to terrorist attacks scenario such as accidental releases of CWA or from historic, buried, munitions are also subjects from response planning. VX is one of the most difficult CWA to destroy. In general, the decontamination of VX can be achieved either by hydrolysis or by oxidation. However, hydrolysis of VX leads to toxic hydrolyzed products. Therefore, decontamination of VX by oxidation is the preferred method. VX contain bivalent sulfur atoms that can be readily oxidized.

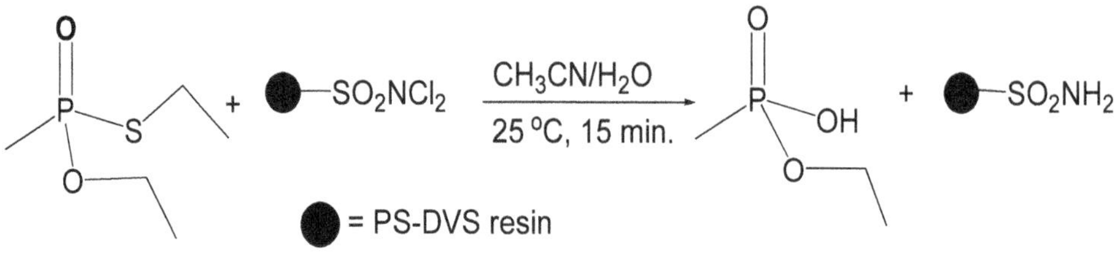

Scheme. 5. Decontamination of OSDEMP with polymeric Beads in aqueous medium

We investigated its use as an alternative reagent for decontamination of simulant of VX at room temperature. To gain more complete understanding of agent chemistry, a simulant of VX, O,S-diethyl methyl phosphonothiolate (OSDEMP) was synthesized initially for decontamination study. The most common and widely used process is the oxidation of VX using N-chloro compounds, which decontaminates it with the formation of nontoxic product.

Decontamination studies of OSDEMP (Gutch, P.2011) were carried out at room temperature in aqueous medium using a mixture of acetonitrile and water (5 : 1). Decontaminated product was separated by GC (Fig 6). OSDEMP was decontaminated 100% with N,Ndichloro poly(styrene-co-divinyl benzene) sulfonamide. The decontaminated product was analyzed as their methyl ester derivatives by reacting with freshly prepared diazomethane in ether. The organic layer did not show any peak corresponding to OSDEMP indicating the reaction of N, N-dichloropoly (styrene-co-divinyl benzene) sulfonamide with OSDEMP. The degradation of OSDEMP with N,N-dichloro poly(styrene-co-divinyl benzene) sulfonamide followed by P-S bond cleavage via oxidation and hydrolysis leading to the formation of nontoxic product ethyl methylphosphonate (EMPA).

In conclusion, the study reveals that N, N-dichloro poly (styrene-co-divinyl benzene) sulfonamide works as an excellent decontaminating agent against OSDEMP, which bears oxidizable bivalent sulfur by its oxidation followed by hydrolysis to nontoxic product EMMP in aqueous medium at room temperature. This reagent has advantage over earlier reported reagent in terms of effectiveness, stable, nontoxic, cheap, easy to synthesize, recyclability (collected after filtration, rechlorinated, and used for further reaction), and decontamination of simulant of VX to give nontoxic product EMPA at room temperature.

N, N-dichloro poly (styrene-co-divinyl benzene) sulphonamide is a efficient reagent to fulfil different required objective to organic synthesis. The main advantage of this reagent is that reactions were clean and operationally simple.

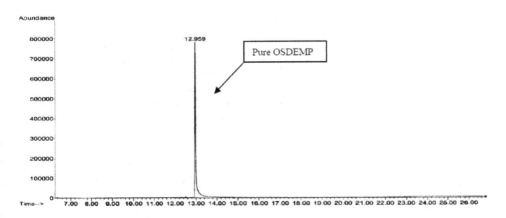

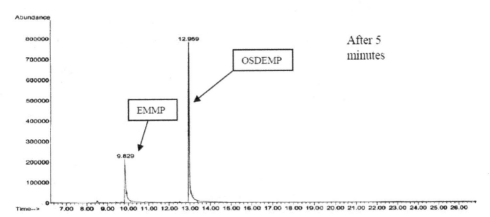

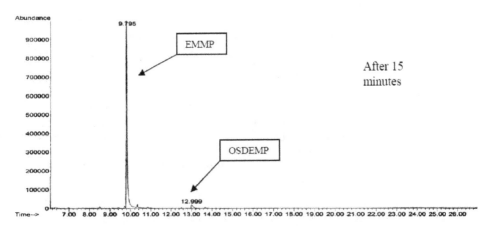

Fig. 6. Progress of a decontamination reaction of OSDEMP as monitored by GC-MS analysis at different time intervals

4. Identification of tear gas compounds in air

Chemical warfare agents can be classified into two general categories, those that exert a lethal effect and those that act in an incapacitating manner. Lethal chemical warfare agents include nerve agents such as Sarin, Soman and Tabun, while incapacitating agents include irritants (tear gases or riot control agents). Acute exposure to irritants causes a number of incapacitating effects including burning or irritation of the skin and eyes, coughing, nausea and vomiting. The incapacitating nature of these chemicals has led to the development of dispersal devices for their use in riot control situations, during military training exercises and to a lesser extent as chemical weapons on the battlefield. The most commonly employed irritants (Malhotra, R. 1987) are o-Chlorobenzylidenemalononitrile, (Gutch, P 2005) often referred to as tear gas, and 2-chloroacetophenone. Dibenz [b,f]-l,4-oxazepin has been used less frequently and 1-methoxycycloheptatriene was evaluated as a possible military training agent.

The text of the 1997 "Convention on the Prohibition of the Development, Production, Stockpiling and Use of Chemical Weapons and their Destruction" states in Article 1 that: "Each state party undertakes not to use riot control agents as a method of warfare". United Nations peacekeeping forces could encounter use of irritants during active duty in regions of the world where there is a threat of chemical warfare agent use. Intelligence gathering, through the collection of contaminated samples, and subsequent analysis of the samples would enable identification of the suspect chemical and confirm use of a controlled chemical for warfare purposes. The results of such analyses would likely contribute to the development of appropriate strategic and political positions. Gas chromatographic (GC) methods, including methods based on GC retention indices, have been used for the detection of irritants in suspect samples.

Methods for detection, identification and quantitative determination of various tear gas compounds such as Dibenz [b,f]-l,4-oxazepin (CR) (Gutch, P (2007), ω-Chloroacetophenone (CN) (Nigam, A. 2010), 2-Chlorobenzylidene malononitrile (CS) and their analogues are required during their production and also for verification of their use for prohibited activities under Organisation of prohibition of chemical weapons (www.opcw.nl)(**OPCW**). The availability of identification data on CS, CR, CN and related compounds would facilities the verification of in case of alleged use of these chemicals. United Nations peacekeeping forces could encounter use of irritants during active duty in regions of the world where there is a threat of chemical warfare agent use. Intelligence gathering, through the collection of contaminated samples, and subsequent analysis of the samples would enable identification of the suspect chemical and confirm use of a controlled chemical for warfare purposes. The results of such analyses would likely contribute to the development of appropriate strategic and political positions. Gas chromatographic (GC) methods, including methods based on GC retention indices, have been used for the detection of irritants in suspect samples. The correlation of retention indices between an unknown and reference compound on two or more columns of different polarities is generally sufficient for identification purposes. For the evolution of retention indices, a homologues series of n-alkanes is commonly used as a reference compound.

4.1 Measurement of retention indices

In order to measure the retention indices generally solutions containing several tear gases in acetone together with n-alkanes standards were injected onto the GLC column, the sample size in each instance was about 0.1 ml. The amounts of individual compounds and n-alkanes present in the range 10-15 mg. The retention times (Fig.7) were recorded with the accuracy of up to 0.01 min with the help of Shimazda CR3A data processor. In order to identify individual components in the mixture, an authentic sample of each tear gas was injected separately and its retention time was compared with that of the component in mixture. Temperature-programmed retention indices for individual compounds was calculated by using equation 1. The measurement of retention indices in two columns with different polarities is shown to be applicable for identification of tear gases compound in air.

4.2 Determination of retention indices under programmed temperature conditions

Programmed temperature GC allowed the analysis of a number of compounds over a wide range of volatilities in a single run. Under programmed temperature conditions, a linear relationship exists between the retention time of n-alkanes and their carbon number. Hence, under these conditions, it is possible to calculate retention index value using retention time only. The retention indices under programmed temperature chromatographic condition (RI_p) were calculated using the van den Dool and Kratz formula. (Equation 1).

$$RI_p = 100\left(\frac{t_c - t_z}{t_{z+1} - t_z}\right) + 100z \qquad (1)$$

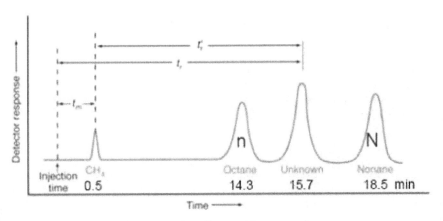

Fig. 7. GC chromatogram of mixture of n-alkane with tear gas.

Here t_c, t_z and t_{z+1} were retention times of the solute, alkane eluted immediately prior to the compound with z number of carbon atoms (lower alkane), and alkane eluted immediately after the compound with z+1 number of carbon atoms (higher alkane), respectively.

4.3 Determination of retention indices under isothermal chromatographic conditions

Under isothermal conditions unlike programmed temperature conditions, there exists a non-linear relationship between retention time and number of carbon atoms of n-alkanes. Therefore, for the calculation of retention indices under these conditions, logarithm of the corrected retention time was taken into account. Retention indices under isothermal experimental conditions (RI_1) were calculated using the Kovats formula (Equation.2)

$$RI_I = 100\left(\frac{\log t_c' - \log t_z'}{\log t_{z+1}' - \log t'}\right) + 100z \tag{2}$$

Here t'_c, t'_z and t'_{z+1} were corrected retention times of the solute, alkane eluted immediately prior to the compound with z number of carbon atoms (lower alkane), and alkane eluted immediately after the compound with z+1 number of carbon atoms (higher alkanes), respectively.

We have reported temperature programmed retention indices for several tear gas compound using Van den Dool's equation. GC of these compounds as a class on a fused silica capillary column and temperature –programmed retention indices for most of them were not previously reported. Further the measurement of retention indices on two columns with different polarities is shown to be applicable for the identification of tear gases in air. We have reported temperature programmed retention indices (Gandhe, B. 1989). on various tear gas compounds such as Methyl ethyl ketone, Chloro acetone, Bromo acetone, Ethyl bromo acetate, Benzoyl chloride, ω-Chloroacetophenone (CN), and 2-Chloro benzylinedinemalononitrile (CS) on polar DB-1 and non polar DB-1701 fused silica capillary column.

Riot control agent2-Chlorobenzylidenemalononitrile (CS) and its analogues have both skin irritating and lacrymating properties. We have also reported retention indices (RI) of CS and its thirteen analogues (Gutch, P 2004) relative to the homologues n-alkanes series. These values are determined on nonpolar BP-1 and polar BP-10 capillary column under programmed temperature and isothermal chromatographic condition. The analogues differ in substitution at ortho or para position of phenyl ring and retention indices are found to vary according to the nature of the substituent.

5. Conclusion

Gas chromatography has been successfully applied to the separation of mixtures of numerous organic and inorganic compounds that have appreciable vapour pressure. Gas chromatography is also used for analysis of pollutants, analysis of alcoholic beverages, analysis of pharmaceutical and drugs, analysis of clinical applications, analysis of essential oils, analysis of Fatty acids, analysis of explosives, and analysis of pesticides. Gas chromatography is very useful analytical technique for identification of known and unknown volatile and thermo labile compounds and monitoring of various organic reactions. Gas chromatography with a fused silica capillary column is the method of choice due to its high resolution and sensitivity.

6. Acknowledgment

We thank Dr. R. Vijayaraghavan, Director, DRDE, Gwalior, for providing necessary facilities and for useful discussion.

7. References

Akin, Akdag. (2006). Oxidation of thiols to disulfides with monochloo poly (styrenehydantoin) beads. *Tetrahedron letter*, Vol. 47, pp. 3509-3510.

Bahman, Tamani. (2009). Synthesis and application of cross-linked Poly (N-bromomaleimide) in oxidation of various organic compounds. *Iranian. Polymer. Journal*, Vol. 18, 12, pp. 957-967.

Bogoczek, R. (1989). Studies on a macromolecular dichloramine: The N, N-dichloro poly (styrene-co-divinyl benzene) sulfonamide. *Angew Makrolmolec Chem*, Vol. 169, pp. 119- .

Dubey, D. (1999). Reaction of bis (2- chloroethyl) sulphide with N,N′-dichlorobis (2,4,6-trichlorophenyl) urea. *J Org Chem*, Vol. 64, pp. 8031-8034.

Eto, M. (1974)*Organoo phosphorous pestisides organic and biological chemistry*. CRC press, USA.

Gandhe, B. (1989). Gas chromatographic retention indices of tear gases on capillary columns. *Journal of Chromatography*, Vol. 479, pp. 165-169.

Ghorbani, Vaghei. (2009). Poly (N, N′-dichloro-N-ethylbengene-1, 3-disulfonamide) and N, N, N′, N′-tetra-chlorobengene. *Synhtesis*, Vol. 6, pp. 945-950.

Gupta, Hemendra. (2008). N,N-Dichloro poly (styrene-co-divinyl benzene) sulfonamide polymeric beads: an efficient and recyclable reagent for the synthesis of dialkyl Chlorophosphates from dialkylphosphites at room temperature. *Tetrahedron Letters*, Vol. 49, pp. 6704–6706.

Gutch, P (2004). Chromatographic analysis of BMN analogues of 2-chloro bezylidenemalononitriles. *J Indian Chem.Soc*, Vol. 81, pp. 874-878.

Gutch, P (2005). Structure and biological activity relationship of 2-chloro bezylidenemalononitriles, A Riot Control agent. *Defence Science Journal*, Vol. 55, pp. 447-457.

Gutch, P (2007). A Simple, Convenient And Effective Method For the Synthesis of Dibenz (b,f) 1,4-Oxazepines (CR); A New Generation Riot Control Agent And Its Analogues. *Heterocyclic Communication*, Vol. 13, pp. 339- 396,

Gutch, P. (1997). Studies on the synthesis, structure-activity relationship and degradation of substituted malononitriles. Ph.D Thesis , Jiwaji University Gwalior.

Gutch, P. (2007). N,N-Dichloro Poly(styrene-co-divinyl benzene) Sulfonamide–Synthesis, Characterization and Efficacy against Simulant of Sulfur Mustard. *J. Applied Polymer Science*, Vol. 105, pp. 2203-2207.

Gutch, P. (2008). Polymeric decontaminant 2 (N,N-dichloropolystyrene sulfonamide): Synthesis, characterization, and efficacy against simulant of sulfur mustard. *J. Applied Polymer Science*, Vol. 107, pp. 4109-4115.

Gutch, P. (2011). New Effective Synthetic Method for Preparation of N,N-Dichlorocarbamates. *Synthetic Communications,* Vol. 41, pp. 1554-1557,

Gutch, P. (March 2001). Synthesis, Characterization and Mass Spectrometric fragmentation of gem. Dicyano oxidants derived from Benzilidene malononitriles. *Indian Journal of Chemistry*, Part B, Vol. 40, B, pp. 243-247.

Gutch, P. (2010). N, N-Dichloro poly (styrene-co-divinyl benzene) sulphonamide polymeric beads: An efficient and recyclable decontaminating reagent for Sulfur Mustard. Proceeding in Macro 2010 at IIT Delhi. pp. 15-17.

Gutch, P. (2011). N, N di chloro poly (styrene-co-divinyl benzene) sulphonamide polymeric beads An Efficient recyclable decontaminating reagent for O,S diethyl methyl phosphonothiolate a similuent of VX. *J. Applied Polymer Science*, Vol. 121, pp. 2250-2256.

Gutch, P. (2011). N, N-dichloro poly (styrene-co-divinylbenzene) sulfonamide beads as an efficient, selective and reusable reagent for oxidation of thiols to disulfides. *International J. Polymeric Materials*, Communicated.

Gutch, P. (2011). Polymeric Decontaminant: N, N - Dichloro 2, 6 - Dimethyl 1, 4 - Phenylene Oxide Sulphonamide: Synthesis, Characterization & Efficacy against Simulants of Sulfur Mustard. J. Polymer Materials, Vol. 28, 1, pp. 5-13.

Jones, G. (1972). CS and its chemical relatives. *Nature*, Vol. 235, pp. 257-259.

Kolvari, E. (2007). Application of N-halo reagents in organic synthesis. J. Iran. Chem. Soc, Vol. 4, 2, pp. 126-174.

Koval, I. (2002). N-halo reagent: N-halosuccinimide in organic synthesis and in chemistry, a natural compound. *J Iran Chem Soc*, Vol. 38, pp. 301-337.

Koval, I. (2001). N-Halo Reagents. Synthesis and Reactions of N-Halocarboxamides. *Russ J Org Chem*, Vol. 37, 3, pp. 297-317.

Kowalski, P. (2005). Oxidation of sulphides to sulfoxide, Part 1: Oxidation using halogen derravitives. *Tetrahedron*, Vol. 61, pp. 1933-1953.

Malhotra, R (1987). Chemistry and toxicity of tear gases. *Def. Sci. J*, Vol. 37, pp. 281-296,

Nigam, A. (2010). Thermal decomposition studies of riot control agent Chloro acetophenon (CN) by Pyrolysis-gas chromatography-mass spectrometry. J. of Hazardous Materials. Vol. 184, pp. 506-514.

(1977). *Convention on the Probhition of the development production stockpiling and use of Chemical weapons and on their destruction, technical secretariat of the organization for prohibition of chemical weapons to The Hague*, Available from: < www.opcw.nl.> Accessed 28 May 2008.

Pathak, U. (2009). Efficient and convenient oxidation of thiols to symmetrical disulfide with silica-PCl_5/$NaNO_2$ in water. *Syn. Comm*,Vol. 39, pp. 2923-2927.

Pattssion, J. (1973). *A Programmed Introduction to gas Chromatography.* Spectrum House Alderton Crescent, London, Great Britain,

Prasad, G. (2009). Modified titania nanotubes for decontamination of sulphur mustard. *J. Hazard. Mater*, Vol. 167, pp. 1192- 1197.

Rose, S. (1969). CS – a case for concern.. *New Sci*, Vol. 43, pp. 468-469.

Singh, R. (2011). Decontamination of Toxic chemicals by N-chlorocompound. Ph.D Thesis , Jiwaji University Gwalior.

Somani, S. (Ed). (1992). *Chemical Warfare Agents*, Academic Press, New York,

Talmage, S. (2007). Chemical warfare agent degradation and decontamination. *Curr. Org. Chem*, Vol. 38, pp. 953-959.

Yang, Y. (1995). Chemical reaction for neutralizing chemical warfare agents. *Chem Ind*, Vol. , pp. 334-336.

Yu-chu, Yang. (1992). Decontamination of chemical warfare agents. *Chem. Rev*, Vol. 92, pp. 1729-1743.

6

Gas Chromatograph Applications in Petroleum Hydrocarbon Fluids

Huang Zeng, Fenglou Zou, Eric Lehne, Julian Y. Zuo* and Dan Zhang
Schlumberger DBR Technology Center, Edmonton, AB,
Canada

1. Introduction

1.1 Composition of reservoir hydrocarbon fluids

In the petroleum hydrocarbon fluids, the most commonly found molecules are alkanes (linear or branched paraffins), cycloalkanes (naphthenes), aromatic hydrocarbons, or more complicated compounds like asphaltenes. Under surface pressure and temperature conditions, lighter hydrocarbons such as CH_4, C_2H_6, and inorganic compounds such as N_2, CO_2, and H_2S occur as gases, while pentane and heavier ones are in the form of liquids or solids. However, in petroleum reservoir the proportions of gas, liquid, and solid depend on subsurface conditions and on the phase diagram (envelop) of the petroleum mixture. To obtain compositions of a reservoir fluid, a reservoir sample is flashed into gas and liquid phases at ambient conditions. The volume of the flashed gas, and the mass, molar mass and density of the flashed liquid are measured. Then a gas chromatograph is used to analyze compositions of the gas and liquid phases as described briefly below. The recombined compositions based on the gas and liquid according to the measured gas/oil ratio are those of the reservoir fluid.

Generally speaking, crude oils are made of three major groups:

- Hydrocarbon compounds that are made exclusively from carbon and hydrogen;
- Non-hydrocarbon but still organic compounds that contain, in addition to carbon and hydrogen, heteroatoms including sulfur, nitrogen and oxygen;
- Organometallic compounds: organic compounds, normally molecules of porphyrin type that have a metal atom (Ni, V or Fe) attached to them.

1.2 Hydrocarbons

Hydrocarbons are usually made of few groups:

a. linear (or normal) alkanes (paraffins)
b. branched alkanes (paraffins)
c. cyclic alkanes or cycloparaffins (naphthenes)
d. aromatic alkanes (aromatics)

---

* Corresponding Author

From the GC perspective, the analysis of alkanes is performed using a non-polar column and separation is based on boiling point. Normal alkanes boil few degrees higher than their respective branched ones. In Table 1.1 an example illustrating this point is given.

Compound	Boiling point (ºC)
n-Octane	126
2-methylheptane	116
3-methylheptane	118
4-methylheptane	117

Table 1.1. Boiling points of octane isomers

From the data above, the branched alkanes are closer together and the corresponding normal alkanes boil at higher temperature. Thus, branched alkanes elute first, followed by the normal alkanes. For GC analysis, it is recommended to integrate the end of an alkane to the end of the next alkane as a family of one particular alkane, as shown in Figure 1.1.

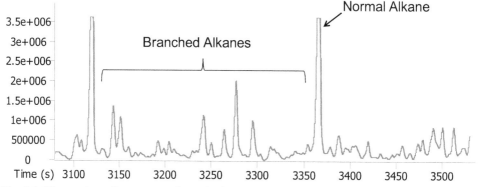

Fig. 1.1. Illustration of integration branched and linear alkanes

In 1873, van der Waals introduced the first cubic equation of state (EOS) by modifying ideal gas law. In 1949, Redlich and Kwong modified the van der Waals EOS which was then modified by Soave (1972). Peng and Robinson (1976) introduced the Peng-Robinson EOS for better liquid volume calculations. Many cubic EOS were developed later. Cubic equations of state such as the Peng-Robinson EOS with volume translation have been widely used for the calculations of fluid phase behaviour for hydrocarbon systems. Based on the recombined composition of the reservoir fluid, the characterization procedure of Zuo and Zhang (2000) can be used to characterize single carbon number (SCN) or true boiling point (TBP) fractions and plus fractions. Then cubic EOS can be employed to calculate phase behaviour of the reservoir fluid. The EOS (compositional model) or simulated fluid properties (black oil model) is used in reservoir simulators such as Eclipse and/or process simulators such as HYSYS. For polar systems, cubic EOS can also be used by coupling complicated mixing rules such as the Huron-Vidal mixing rule and the Wong-Sandler mixing rule. On the other hand, Davarnejad et al. (2007, 2008) considered the Regular Solution Equations as a general model for polar and non-polar systems.

1.3 Non-hydrocarbons

The hydrocarbons that contain heteroatoms could vary from very simple one such as thiophene to very complex mixtures such as asphaltenes for which the structure is not well understood, but known to contain sulfur, oxygen and nitrogen at different levels, in addition to carbon and hydrogen (Buenrostro-Gonzalez *et al.*, 2002; Woods *et al.*, 2008).

The most common method to separate petroleum fractions is called SARA, which stands for saturates, aromatics, resins and asphaltenes. It needs to be noted that cyclic compound are included in the fraction of saturates. The light fraction is made mostly of alkanes and aromatics. Light aromatics containing heteroatoms could be distilled off with this fraction. The split between alkanes and aromatics could be performed using supercritical fluid chromatography by changing the solvent strength (Dulaurent *et al.*, 2007).

The heavy fraction is first subjected to asphaltenes precipitation using an excess of normal alkanes such as n-pentane or n-heptane, usually at a oil-to-alkane ratio of 1 to 40. Different methods exist in the literature for asphaltenes separation (Kharrat, 2009). After extraction of asphaltenes, the maltenes are separated into three fractions: saturates, aromatics and resins using solvents with increasing polarity as indicated in Figure 1.2.

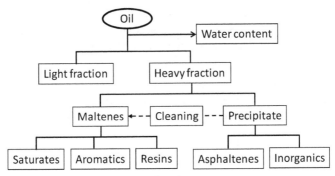

Fig. 1.2. Chart flow of fractionation of a crude oil

The saturate fraction is analyzed by gas chromatography, leading to n-alkanes content. Aromatics are analyzed by Gas Chromatography with Mass Spectrometry (GC-MS). Resins and asphaltenes are the most difficult to analyzed by GC because of their high boiling points. Therefore, the applications of GC on the analysis of heavy oil, which has a high concentration of asphaltene and resin fractions, are limited. In a reported high temperature GC (HTGC) technique, a short 5-m glass capillary column was used to elute compounds in bitumen and bitumen-derived products with boiling points as high as 700°C (equivalent to alkane with carbon number of 90, C_{90}) (Subramanian *et al.*, 1996).

In this chapter, the principles and instrumentations of several GC techniques, and their applications on the analysis of petroleum hydrocarbon fluids are reviewed.

2. High Temperature Gas Chromatography (HTGC)

GC has the advantages of high column efficiency, high sensitivity, fast analysis speed and ease to be combined with other analytical methods (e.g. Mass Spectrometry). Thus, it is widely used to analyze crude oil and its products. Because of the limited thermal stabilities of capillary column and the stationary phase, the maximal column temperature of conventional GC is around 325°C, and the analysis is limited to hydrocarbons with carbon number less than about 35. This fact limits the GC applications on the analysis of alkanes of high carbon numbers (C_{40+}) which are important for some areas including organic geochemistry.

During the past decades, high temperature (325-450°C) GC (HTGC) has been developed rapidly and used for components of high molecular weight in crude oil, *etc.*

2.1 Requirements on the parts of HTGC

The main materials for the HTGC column include stainless steel and fused silica. Steel column has excellent mechanical properties at high temperature, but it has very strong catalytic and adsorptive effects. Therefore, a deactivation inner coating between the steel tubing and the stationary phase is needed. Another drawback of stainless steel column is that it difficult to cut. A protective coating is needed on the outside of the fused silica column to maintain the flexibility of the column. Right now the frequently used coating materials include polyimide and aluminum (Kaal & Janssen, 2008). It has been suggested that the polyimide may be broken down above 360°C and alumina coating can overcome this problem. However, the alumina coating on silica column can become brittle upon repeated heating above 400°C (Application note #59551, Restek Corporation).

The stationary phase also needs to be stable above 400°C upon repeated heating with minimal breakdown. It is mostly based on highly thermostable polysiloxane which can be bonded onto the capillary inner wall via the condensation reaction between the silanol terminal groups of the polysiloxane and the silanol groups on the silica surface during curling process (Mayer *et al.*, 2003; Takayama *et al.*, 1990). The commonly used materials for stationary phase of HTGC include carborane-siloxane polymers (maximum temperature up to 480°C) and silphenylene-modified polysiloxane (maximum temperature up to 430°C), *etc.* (Kaal & Janssen, 2008).

Injection method is very important for HTGC. Cold on-column injection is preferred because of its ability to eliminate discrimination against the most non-volatile compounds (Damasceno *et al.*, 1992). For many compounds with high boiling points, programmed-temperature vapourisation injection (PTV) also gives good results (van Lieshout *et al.*, 1996).

The most frequently used detection method for HTGC is flame ionization detection (FID). Other detection methods have also been used, including mass spectrometery (MS) (Hsieh *et al.*, 2000; Philp, 1994), atomic emission detection (AED) (Asmussen & Stan, 1998) and inductively coupled plasma mass spectrometry (ICP-MS) (Glindemann *et al.*, 2002), *etc.*

2.2 Applications of HTGC

2.2.1 Simulated distillation of crude oil

The crude oil is composed of a large amount of alkanes with different carbon numbers, giving rise to a broad range of boiling point. The understanding of the carbon number distribution of crude oil can help to precisely evaluate the factors affecting the properties of crude oil and the oil products. It is also important for the designing of the distillation, processing equipment and the quality control of the products.

Normal true boiling point (TBP) distillation involves a long procedure and is costly, and the distillation temperature is normally limited even when vacuum is used. GC has been used widely as a fast and reproducible method for simulated distillation (SimDis) to analyze the carbon number distribution of the hydrocarbon in the crude oil. When HTGC is used, SimDis method can reach a boiling point range of 35-750°C, equivalent to n-alkanes with a carbon number distribution of C_5 to C_{120} (Kaal & Janssen, 2008). HTGC SimDis normally uses a short capillary column with a thin film of polydimethylsiloxane stationary phase. Recently, Boczkaj et al reported the possibility of using an empty deactivated fused silica column (EC-GC) for HTGC-SimDis (Boczkaj et al., 2011).

Figure 2.1 presents an example of high temperature simulated distilation (HTSD) results for heavy and light oils. The heavy oil starts distillation at a higher temperature (~200°C) than for the lighter oil (~150°C). The residue, which is the fraction that does not distill at 700°C, is much higher for the heavy oil than for the lighter one.

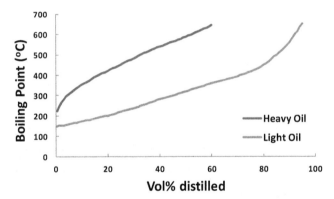

Fig. 2.1. Example of HTSD results for a heavy (red) and light oil (blue).

2.2.2 Wax analysis

Waxes are solids made up of heavy hydrocarbon (C_{18+}) which are mainly normal alkanes (paraffins) (Kelland, 2009, p. 261). The waxy oil can be used to produce wax-based products, and normally has low concentrations of sulphur and metal which are harmful for refinery. But on the other hand, when temperature of the crude oil drops during oil production, transportation or storage, paraffin waxes in the crude oil can precipitate and make serious problems including pipeline blockage and oil gelling, etc. Thus, it is important to measure the composition (amount and type) of wax in the crude oil, and to estimate the temperature at which the wax will crystallize (wax appearance temperature, WAT) and the wax precipitation curve (WPC) to understand the potential wax problem and its magnitude.

Compared with conventional GC, HTGC significantly extends the range of detectable hydrocarbon. Therefore, HTGC has become more and more routinely used for wax analysis, and the HTGC results can be correlated to the physical properties of the wax, including melting point, refractive index and kinematic viscosity, etc (Gupta & Severin, 1997).

Currently, there are no standard wax-analysis methods, but some methods are developed by the petroleum industry. These methods remain proprietary. Figure 2.2 shows a typical HTGC chromatogram of a waxy crude oil. With a calibrated column, the relative area under the curve for each n-alkane will be converted to the relative abundance of the species. The reported n-alkane composition will be the amount of each n-alkane chain number and relative or absolute abundance (see an example as shown in Figure 2.3).

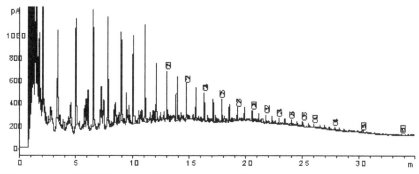

Fig. 2.2. Examples of HTGC traces of a waxy crude oil

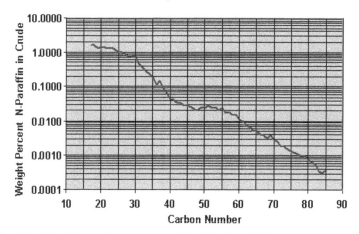

Fig. 2.3. n-Paraffin distribution for a waxy crude oil analyzed by HTGC

An important application of HTGC analysis on waxy crude oils is to measure high molecular weight hydrocarbons (HMWHCs) which provide desirable geochemical information, and some significant findings have been reported.

Del Rio and Philp reported HTGC analysis of some wax samples blocking oil wells and found that the wax deposits were normally composed of hydrocarbons with maximal carbon number around 40 to 50 (del Rio & Philp, 1992). Roehner *et al* used an extended HTGC method to determine the compositions of crude oil solids and waxes up to C_{60} formed in the Trans-Alaska Pipeline. A longer capillary column was used to achieve an improved resolution of high carbon number groups. The n-alkane/non-n-alkane ratio was used to distinguish between certain types of crude oil solids (Roehner *et al.*, 2002).

HTGC has been used to study/monitor the biodegradation of crude oils. For example, using a quantitative HTGC method, Heath *et al* estimated the biodegradation of the aliphatic fraction of a waxy oil. It was found that light hydrocarbon was quickly biodegraded and HMWHCs (C_{40+}) showed resistance to biodegradation (Heath *et al.*, 1997).

HTGC is also an useful tool to study the origin/source of the crude oils. Hsieh and Philp measured HMWHCs from crude oils derived from different sources, including terrigenous, lacustrine, marine source material as well as source rock extracts. Different structural compositions of these HMWHCs (including alkyl-cycloalkanes, methylbranched alkanes, and alkyl-aromatic hydrocarbons) have been revealed by HTGC and it was found that the fraction of the HMWHCs in the whole oil is significantly higher than previously thought (can be up to 8%) (Hsieh & Philp, 2001; Hsieh *et al.*, 2000). The distribution of long chain, branched and alkylcyclohexanes of the HMWHCs analyzed by HTGC has been used as a useful mean to distinguish oils derived from different sources (Huang *et al.*, 2003).

It has been pointed out that the sample extraction/cleaning procedure is crucial to get representative HMWHC samples from crude oil for HTGC measurement (Thanh *et al.*, 1999). More details on the HTGC anslysis of HMWHCs and it applications can be found in an overview given by Philp *et al* (Philp *et al.*, 2004).

WAT and WPC are two important parameters for wax-related flow assurance problems. They can be experimentally measured (Kelland, 2009), and can also be predicted by different thermodynamic models. All these models rely on the experimental data of n-paraffin distribution which is now commonly provided by HTGC. For example, Zuo and Zhang have developed a model to predict the WAT based on the oil composition provided by HTGC (Zuo & Zhang, 2008). The model has been proved to provide prediction results in good agreement with experimental data of synthetic oils and reservoir fluids, and several examples are given in Figures 2.4 to 2.6 (details givein in Zuo & Zhang 2008). Coto *et al* analyzed three parameters to improve the n-paraffin distribution provided by HTGC, including total amount of C_{20+} paraffin, extrapolation of C_{38+} paraffin and molecular weight of the crude oil. The results showed that the distribution of the n-paraffin has great impact on the accuracy of the prediction model (Coto *et al.*, 2011).

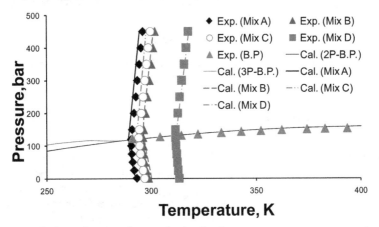

Fig. 2.4. WAT and phase diagram for synthetic oil mixtures.

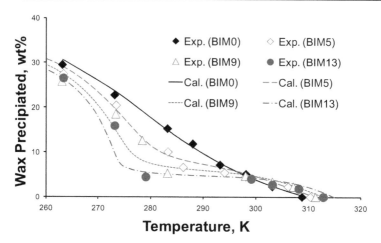

Fig. 2.5. Wax amount for synthetic oil mixtures.

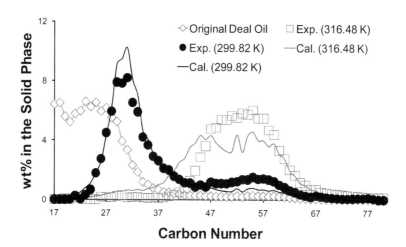

Fig. 2.6. The wax distributions vs. carbon numbers are compared for a crude oil.

3. 2D-GC (GC-GC)

3.1 Concept of comprehensive GC × GC

Since Liu and Phillips depictured comprehensive two dimensional gas chromatography (2D-GC) in 1990s (Liu & Phillips, 1991; Vendeuvre *et al.*, 2007; Zrostlíková *et al.*, 2003), the technology of 2D-GC has developed rapidly and been applied in the areas such as biological/clinical, environmental, food, forensics, petroleum, pharmaceuticals and fragrances in forensic, food and petroleum oil characterization (Wang *et al.*, 2010). 2D-GC employs two capillary GC columns of different selectivity coupled by a modulator (Figure 3.1) which will be introduced here.

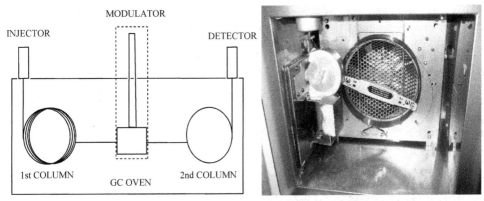

Fig. 3.1. Scheme of comprehensive two dimensional gas chromatography (left). A photo of 2D-GC (right), where the primary column located in the main GC oven, and the second column housed in an independent oven within the main oven.

3.1.1 First column

The first column is typically of conventional length, longer, wider and thicker than the second column, being (15–30) m (long) × (0.25–0.32) mm I.D. (inner diameter) × (0.1–1) μm film (df). The stationary phase of the column can be non-polar or polar. For example, 100% dimethylpolysiloxane is considered as non-polar, and phenyl-substituted methylpolysiloxane is considered as polar column (Scheme 3.1), and the more phenyl contained, the more polar the column is (Betancourt *et al.*, 2009).

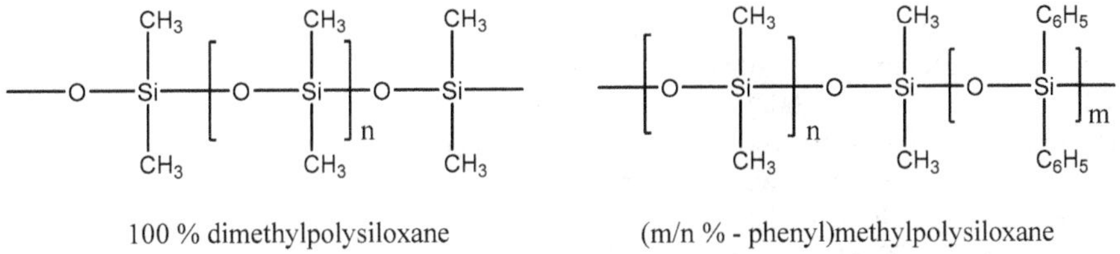

100 % dimethylpolysiloxane   (m/n % - phenyl)methylpolysiloxane

Scheme 3.1. Structures of dimethylpolysiloxane and phenyl-substituted methylpolysiloxane

3.1.2 Second column

The second column is shorter, narrower and thinner, of (0.5–2) m × 0.1mm I.D. × 0.1μm df. It is usually more polar or less polar than the first column. If the second column is more polar than the first column, it is called non-polar × polar configuration; if the second column is less polar than the first one, it is called polar × non-polar configuration. In most time, 2D-GC has non-polar × polar configuration. The two columns can be housed in the same oven or in separate ovens to enable more flexible temperature control (Li *et al.*, 2008).

3.1.3 Modulator

A modulation unit, placed between the two columns, is the most critical component of 2D-GC (Adahchour *et al.*, 2008; Adahchour *et al.*, 2006; Marriott *et al.*, 2003; Pursch *et al.*, 2002;

Wang & Walters, 2007). Modulator periodically samples effluent from the first column and injects it into the second column. After the second-dimension separation is completed the next modulation starts. In this way, all compounds are subjected to two different column separation mechanisms, and are resolved based on two different aspects of chemistry, or selectivity. Each modulation period is so short that each peak from the first column is cut into several smaller slices that go through separation on the second column. Therefore, GC × GC chromatography peaks become much narrower compared with traditional 1D GC peaks. The setup of two columns (the first column is longer and wider and the second column is shorter and narrower) ensures that the total second dimension (2D) separation is completed in the run time of the first-dimension analysis.

It should be noted that comprehensive GC × GC is different from two-dimensional "heart-cut" gas chromatography (Zrostlíková *et al.*, 2003) in which one zone of effluent eluting from the first column is isolated and subsequently separated in a different column.

Modulator is also a reference timing signal at the interface between the two columns (Phillips & Beens, 1999). There are different kinds of modulators such as valve modulator, thermo-modulator and cryogenic modulator, which are very well reviewed (Betancourt *et al.*, 2009; Phillips & Beens, 1999). Nowadays most of modulators are dual-stage cryogenic modulator using CO_2 or liquid nitrogen as cooling agent. Two-stage cryogenic modulator system has two hot points and two cold points; hot point is used to release and reject the slice from the first column, and cold point is used to hold the slice. For the analyses of whole petroleum oil, liquid nitrogen is a better choice because of the existence of very low boiling point compounds in the light ends of the petroleum oils.

3.1.4 Detector

The dimension of the second column is such that eluting peaks have peak widths in the order of 10-100 ms (Adahchour *et al.*, 2008, 2006; Marriott *et al.*, 2003; Phillips & Beens, 1999; Pursch *et al.*, 2002; Wang & Walters, 2007). To properly sample the narrowest peaks, detectors need to have fast response. Therefore, the sampling rate at which the detector signal is sampled should be at least 100 Hz, but a slower data rate of 50 Hz can be used for wider peaks. Flame ionization detectors, FID, which has negligible internal volumes and can acquire data at frequencies of 50–300 Hz, are most widely used. Mass spectrometer (MS) can provide structural information, enable unambiguous identification, and ensure high selectivity throughout the chromatogram, and hence is a good detection method. Quadrupole mass spectrometers operating in full scan mode are too slow to properly sample a GC×GC peak unless that peak is broadened. Fast time-of-flight mass spectrometers (TOFMS) that operate with spectral acquisition rates of 100-200 Hz are well-suited for GC × GC and have been used for numerous studies. On the other hand, element-specific detectors such as sulphur, nitrogen chemiluminescence detections (SCD, NCD), have been used for nitrogen containing and sulphur containing compounds, respectively (Dutriez *et al.*, 2011; Zrostlíková *et al.*, 2003).

3.2 2D-GC results

3.2.1 GC×GC chromatogram

2D-GC chromatograms can be visualized in traditional 1D version, 2D version (contour plot) and 3D image (surface plot), as shown in Figure 3.2. The contour plot of GC×GC chromatogram conventionally demonstrates the advantage of 2D-GC: structured chromatogram. It clearly shows the different group types in certain patch of the 2D plane, for example, tri-aromatics, bi-aromatic, mono-aromatic, paraffins, hopanes, and so on. Within the ring of biaromatics (naphthalene), different methyl-substitute also distribute orderly, from left to right within the ring being: non-methylated biaromatics (naphthalene), mono-methylated naphthalene, bi-methylated naphthalene, tri-methylated naphthalene and tetra-methylated naphthalene. The GC×GC contour plot makes group-type analysis a great advantage for GC×GC analysis, but it should be mentioned that there are some crossing. For instance, nonylbenzene appears in the area of cyclic group, and some bi-cyclic appears in the area of mono-aromatics. The surface plot of a GC×GC chromatogram against a 1D version of the GC×GC chromatogram clearly shows that 2D-GC has better separation: the peaks crowded in 1D are well separated along the second dimension of the 2D plane, and it also demonstrates other advantages of 2D-GC: high sensitivity and bigger peak capacity. Some peaks are invisible in 1D chromatogram but visible in 2D-GC chromatogram.

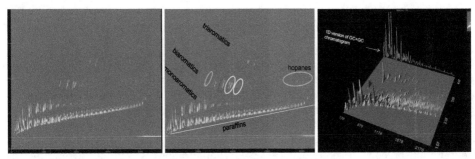

Fig. 3.2. GC×GC chromatograms, first column: VF-1ms 30m × 0.25mm × 0.1μm, second column: BPX50, 1m × 0.1mm × 0.1μm, modulator: 40°C offset to the primary oven, second column: 15°C offset to the primary column. left: Contour plot of a GC×GC chromatogram; middle: contour plot with labelled groups, right: surface plot with the show of 1D version of of GC×GC chromatogram.

3.2.2 Advantages of GC×GC

As demonstrated in the two dimensional contour plot and three dimensional surface plot, it is obvious that GC×GC has the following advantages compared with conventional one demensional GC:

1. Structured chromatograms: the compounds are distributed in the 2D-GC plan according to their group types and each certain group has a certain pattern. Ordered chromatograms have the potential advantage of being much more interpretable than disordered ones. The pattern of peak placement is itself informative and may make it possible in many mixtures to identify most or all of the components or at least to recognize the mixture with good reliability (Adahchour et al., 2008; Li et al., 2008; Phillips & Beens, 1999).
2. Better separation: 2D-GC separates components along the primary dimension and also along the second dimension. Sometimes compounds co-eluted with conventional 1D GC technology can be separated by 2D-GC along the second dimension.

3. Larger capacity: 2D-GC peaks are distributed in the whole plane rather than in one line. 2D-GC has a peak capacity of n1× n2, in which n1 is the capacity along the first dimension and n2 the capacity along the second dimension. Therefore, the capacity of 2D-GC is higher than that of conventional 1D GC.
4. Higher sensitivity: Compared with conventional 1D GC, the sensitivity of 2D-GC is increased by 1.5 – 50 fold (Zrostlíková *et al.*, 2003). Trace amount of analytes can be detected with 2D-GC. The detection limit for 2D-GC is about 2pg (Zrostlíková *et al.*, 2003).

All these characteristics make GC × GC a particularly useful technique for analyzing complex mixture, for example, petroleum oil. The ordered distribution of effluent has been used for quick screening of oil, recognizing the difference between individual oils (Li *et al.*, 2008).

On the other hand, 2D-GC also has some disadvantages. Cooling in cryogenic modulator causes peak tailing along the second column. 2D-GC files are extremely large, and not easy to be applied into different software.

3.2.3 GC × GC column configuration

Normal GC×GC column configuration is from non-polar to polar (the first column is non-polar, and the second column is polar), whereby the sample is separated on the first column based on the boiling point differential of all components, and then further separated on the second column based on the polarity differential. The choice of columns depends on the stationary phase (mainly on polarity), the application temperature of the column, column length and diameter, commercial availability and so on. The normal column configuration is usually good for selecting out high polar components in the samples. Nevertheless, column configuration of polar to non-polar (reversed configuration) is reported to improve the resolution of individual alkanes, cyclokane, branched alkanes, and isoprenoids (Vendeuvre *et al.*, 2005).

Figure 3.3 demonstrates a contour plot of 2D-GC-MS, using a reversed column configuration, from polar (30m×0.25mm×0.25um, DB-17) to non-polar column (1m×0.1 mm×0.1um,

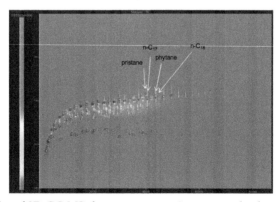

Fig. 3.3. Contour plot of 2D-GC-MS chromatogram using reversed column configuration. First column: polar DB-17 30m×0.25mm×0.25µm; second column: non-polar DB-1MS 1m×0.1 mm×0.1 µm.

DB-1MS). Overall, the contour plot from the reversed configuration looks just like a reversed chromatogram from normal configuration. In the chromatogram of reversed configuration non-polar compounds, like paraffin, have longer second-retention time, and polar compounds have shorter second retention time. In other words, the shorter the second retention time is, the more polar the compounds are. With a reversed column configuration, the compounds are also separated based on boiling point along the first dimensional retention time. For example, with the increase of retention time, the carbon number of normal paraffins increased. However, pristane and phytane come shortly before nC_{17} and nC_{18}, respectively. This is different from the normal column configuration, where pristine and phytane come shortly after nC_{17} and nC_{18}, respectively.

3.2.4 Data processing

GC×GC presents information technology challenges in data handling, visualization, processing, analysis, and reporting due to the quantity and complexity of GC × GC data. Usually, two different softwares are used for data acquisition and data processing, respectively. GC Image (GCImage LLC) is a software system developed at the University of Nebraska-Lincoln that uses advanced information technologies to process and visualized GC × GC data, detect peaks, compare chromatograms, and perform peak deconvolution, pattern recognition and other data mining tasks. ChromaTOF (Leco Corporation) is another software program designed to control Leco's commercially available GC × GC-TOFMS system that has similar functions. Both GCImage and ChromaTOF make effective use of the tremendous amount of data generated when a time-of-flight mass spectrometer is used as a GC × GC detector, including spectral library matching and extracted ion chromatograms. HyperChrom (Thermo Electron Corporation) is a third software program designed to control Thermo Electron's commercially available GC × GC system with flame ionisation detection and employs various data processing and visualization capabilities.

Actually data post process to re-read/re-display the chromatogram in the need of particular application is important, and is proven to be troublesome with the problem of peak alignment, retention time deviation, *etc.* Some particular work has to be done to re-display the chromatogram properly (Aguiar *et al.*, 2011).

3.2.5 Factors affecting 2D-GC retention time and separation

The factors which affect conventional 1D GC retention time and separation all apply to 2D-GC analysis, including:

- inlet temperature
- column temperature and temperature program
- carrier gas and carrier gas flow rates
- the column's stationary phase
- column diameter and length
- sample size and injection technique

Inlet temperature should be set high enough to vaporize injected samples, but not so high that the injected samples can be decomposed (Juyal *et al.*, 2011). Column temperature and temperature program are very important factors affecting GC retention time and separation.

The setup of column temperature and temperature program is to make sure that all of analytes are eluted and separated well. Higher column temperature causes shorter retention times; slower temperature ramp usually leads to better separation. The higher speed of carry-gas results in shorter retention time, which potentially causes peak co-elution. If the amount/concentration of injected sample is too big/high, the chromatogram will look overwhelming crowded with no good separation, especially when the injected samples are crude oil samples. In this case, split injection technique can be employed. When the split ratio is too high (200), the injection accuracy will decrease. Considering the high complexity of crude oil samples, temperature programmed injection technique might be very useful, whereby the sample is introduced in the injector at low temperature and then vaporized by a fast programmed heating process, during which time the split is open all the time and the sample amount entering the column is proportional to the pre-set split ratio (Wang *et al.*, 2010).

Due to the characteristics of 2D-GC, there are several more factors affecting 2D-GC separation, including the modulation period (the second column time), hot pulse time and second column temperature. A short modulation period usually means a short slice from the first column for further separation on the second column, which implies a better separation on the second column. If the modulation period is too long, the separated effluents on the first dimension are accumulated while waiting for injection onto the second dimension, which intends to lose the resolution of the first dimension. If modulation period is too short, on the other hand, peaks wrap-up is unavoidable, resulting in wrong retention times and bad separation.

Currently, the update of 2D-GC operation platform focuses on multi-stage temperature ramp of the second column which will resolve the dilemma between the modulation period and the second column separation time (Betancourt *et al.*, 2009). In some cases, *e.g.* the detection of nitrogen-containing compounds and the analysis on vacuum gas oil (VGO), high modulation period (20s – 30s) is needed. In this case, wide bore first column with bigger diameter are used to make the first dimension peak wider (Dutriez *et al.*, 2009, 2010, 2011).

In 2D-GC analysis with dual stage cryogenic modulator, hot pulse time and cooling period should be adjusted, especially in petroleum oil analysis. The hot pulse time should be long enough to make sure all of samples are re-injected to the second column, and cooling period should be good to make sure all the compounds with low boiling points are refocused on the second column.

3.3 Applications of comprehensive 2D-GC in oil analysis

2D-GC has been used in high temperature analysis of vacuum gas oil (VGO) (HT-2D-GC) (Dutriez *et al.*, 2009, 2010), nitrogen-containing compounds (2D-GC NCD) (Dutriez *et al.*, 2011), sulphur speciation (2D-GC SCD), middle distillates (Vendeuvre *et al.*, 2005), pyrolysis of petroleum source rocks (Py-2D-GC) (Wang & Walters, 2007), biomarkers (Aguiar *et al.*, 2011; Juyal *et al.*, 2011), *etc.*

3.3.1 Group type analysis of two Tarmat oils

Two oils A and B are isolated by a huge tarmat and no connection between the two oils is known. A question arouse regarding to the two oils: are they different? To answer the question, the two oils were subject to independent group type analysis with 2D-GC FID in house and detail analysis with commercial 1D GC-MS. The whole oils were used for group type 2D-GC FID analysis; the two oils were fractionated before detailed 1D GC-MS analysis. The TIC (total ion chromatography) 2D-GC contour plot of oil A is shown in Figure 3.4. There are two steps in the group type analysis: 1) divide each contour plot of the 2D-GC chromatogram of oil A and oil B into 7 areas, tri-aromatic (3-ring-1), bi-aromatic (2-ring-1), mono-aromatic (1-ring-1), hopane (polycyclic-1), sterane (polycyclic-2), before nC_{17} and after nC_{17} at the same retention times (first dimensional retention time, second dimensional retention time), and 2) compare each area of the oil A with the oil B. The analysis results are listed in Table 3.2. The two oils share the same amount of tri-aromatics, bi-aromatics, mono-aromatics, polycyclic-1(hopane compounds), before nC_{17} and after nC_{17}. No sterane compounds (polycyclic-2) were detected in the two whole oil analyses due to too low concentration.

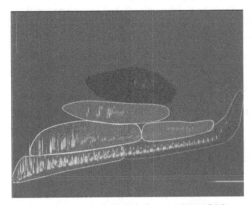

Fig. 3.4. Group type analysis of Tarmat oil A. Dual stage liquid N_2 cryogenic modulator; First column: VF-1MS 30m×0.25mm×0.25μm; second column: BPX50 1m×01.mm×0.1μm. Modulator: 45°C offset to primary oven, second column: 15°C offset to primary column.

The independent 1D GC-MS analyses gave details in hopane, steranes, cheilanthane, adamantanes, naphthalenes, phenanthrenes, benzothiophenes, biphenyls, aromatic steroids and bicyclics. The two oils showed very similar results in each studied items, for example, in biphenyls, as listed in Table 3.3.

So the 2D-GC FID group type analysis of the two oils agrees with the 1D GC-MS detailed analysis on that the two oils are the same.

	Normalized Area%_ Oil A	Normalized Area %_oil B
3-rings	0.41	0.43
2-rings	1.55	1.56
1-ring	7.83	7.86
Polycyclic-1	0.05	0.05
Polycyclic-2	0	0
before nC_{17}	69.53	69.32
after nC_{17}	20.63	20.78

Table 3.2. 2D-GC FID group type analysis of Tarmat oil A and B

	BP	2MBP	DPM	3MBP	4MBP	DBF	DBT
Oil A	7.71	2.68	6.49	8.82	4.82	13.45	253.05
Oil B	8.21	2.88	6.69	8.89	4.93	13.60	252.87

	MBP ratio	DPM ratio	DPM/MBP ratio	DBF ratio	BP/DBT ratio		
Oil A	0.42	0.46	0.40	0.64	0.03		
Oil B	0.42	0.45	0.40	0.62	0.03		

BP: biphenyl; 2MBP: 2-methylbiphenyl; DPM:diphenylmethane; 3MBP:3-methylbiphenyl; 4MBP: 4-methylbiphenyl; DBF: dibenzofuran; DBT: dibenzothiophene; MBP ratio: MBP ratio = 4-methylbiphenyl / (2- + 3-methylbiphenyl); DPM ratio = diphenylmethane / (diphenylmethane + biphenyl); DPM / MBP ratio = diphenylmethane / (diphenylmethane + 2- + 3- + 4-methylbiphenyl); DBF ratio = dibenzofuran / (dibenzofuran + biphenyl); BP / DBT ratio = biphenyl / dibenzothiophene.

Table 3.3. Bipheyl compounds in Tarmat oil A and B based on 1D GC-MS analysis

3.3.2 Advanced product back allocation

Production back allocation of commingled oils from different zones or reservoirs is usually done with 1D GC by comparing composition of commingled oils with involved end members, in which process some subtle differences between inter-paraffin peaks (the GC peaks between n-paraffins) are taken into account. Some uncertainties exist in oil product back allocation when related end members are very similar, or when commingled oils contain heavy oil end members, or the inter-paraffin are not well resolved. GC × GC may help to overcome these uncertainties due to enhanced separation in complex oil. Better identification and quantification of single compounds in heavy oils with GC × GC helps to differentiate the heavy oil end members in commingled oils, and therefore improves heavy oil product back allocation.

In a heavy oil blind production back allocation test, artificial commingled heavy oil #1 (C1) was made of 40% heavy oil A and 60% heavy oil B (C1 = 0.4A +0.6B) , artificial commingled heavy oil #2 (C2) was made of 40% heavy oil B and 60% heavy oil C (C2=0.4B+0.6C), and commingled heavy oil #3 (C3) made of 35% A, 35% B and 30%C (C3=0.35A+0.35B+0.30C). 2D-GC FID analysis was used for allocating all the three samples, of which results was compared with 1D GC FID method. Figure 3.5 shows the 2D-GC FID contour plot of commingled heavy oil C1, where much more peaks are well separated along the second dimension compared with the 1D GC FID. There are two big peaks of CS_2 and toluene, respectively, separated from the other main paraffins in the contour plot, and the peaks are so big that they look like contaminants in the oils. So in the product back allocation, the peaks were omitted to avoid any kind of contamination. The production back allocation results based on 2D-GC FID were compared with that based on 1D GC FID, as shown in Figure 3.6. It shows that overall 2D-GC FID and 1D GC FID show very similar accuracy in the production back allocation of two-end-member commingled samples, but 2D-GC FID shows much high accuracy (error % = 2) than 1D GC FID (error % = 10) in production back allocation of three-end-member commingled sample.

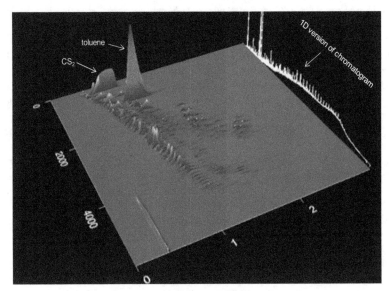

Fig. 3.5. 2D-GC contour plot of commingled heavy oil C1

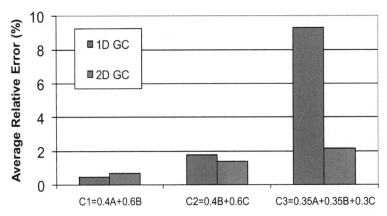

Fig. 3.6. Comparison of heavy oil production back allocation based on 2D-GC FID with 1D GC FID. Ri= real percentage of end members; Ci= calculated percentage of end members. n= number of end members.

4. Gas chromatography fingerprinting

Gas chromatography (GC) analysis is used in the petroleum industry to provide information related to fluid composition, which is needed in petroleum engineering and petroleum geochemistry. Engineering and geochemical approaches, however, require different evaluation techniques for GC chromatograms. Most common application in petroleum engineering is the equation-of-state (EOS) for PVT characterization. Geochemical evaluations include using GC chromatograms for determination of reservoir connectivity, or back allocate commingled production. All these geochemical approaches are based on comparing GC chromatograms on their similarity to each other rather than obtaining compound specific quantitations. This comparison of GC chromatograms is commonly called "geochemical fingerprinting" in the petroleum industry. Fingerprints obtained by gas chromatography (GC) of crude oils is one of the less expensive, less risky and less time consuming methods to study different oils in terms of their similarity to each other. However, fingerprinting using GC chromatograms requires a defined and highly accurate analytical workflow to ensure high precision for assessing the exact similarity index. The analysis and fingerprinting evaluation of reservoir fluids also is affected by uncertainties. Crude oils are complex mixtures containing thousands of different hydrocarbons having huge differences in polarity and molecule size. Consequently, even for a same crude oil sample GC chromatograms can significantly differ when analyzed at different labs or when analyzed using different GC operating conditions. Therefore, it is necessary to keep analytical procedures, the column and the GC operational conditions the same for all GC runs when geochemical fingerprinting is needed. In addition, regular checks on GC performance need to be done to ensure comparability of GC runs. However, a slight shift in retention time is commonly still present when crude oils are analyzed consecutively. The problem mainly results from a slight deterioration in the columns separation performance between consecutive GC runs.

In order to make GC chromatograms directly comparable, it is required to eliminate this retention time shift, which is called warping. Different warping algorithms for retention time alignment of GC chromatograms have been published in the last decade. The most common techniques are dynamic time warping (DTW) and correlation optimized warping (COW) (Nielsen *et al.*, 1998; Tomasi *et al.*, 2004). The comparison of chromatogram similarity index for GC fingerprinting can be done after the preprocessing warping method. The fingerprinting technique relies on comparison of the chemical composition of several chromatograms acquired with the same chromatographic conditions and is based on the differences between peak height ratios of the different crude oil samples. The most advanced and newest technique to determine the similarity index between different chromatograms is the Malcom distribution analysis, recently described in Nouvelle and Coutrot (2010). Malcom distribution analysis uses a statistical method, based on consistent quantification of the uncertainty from chromatography peak height measurements, which provides absolute distances between fingerprints on a universal scale. The method is able to discriminate samples even if the amplitude of the compositional differences is about the same as the error in peak height measurements, and distances between samples are independent of the number of peak ratios available, and the uncertainty in the peak height measurements. The distribution analysis method uses the matrix of all neighboring non-alkane peak height ratios. This method provides a "chemical distance" between each couple of analyses on the basis of the statistical analyses of their respective peak ratios.

In a first step, the distribution histogram of all peak ratio differences available between pairs of chromatograms is built. Then, the inter-quartile range (IQR), as a measurement of the spread of the distributions, is used to characterize the statistical distance between the fingerprints. Since the GC column becomes progressively degraded during analysis, the method requires the peak ratios of two chromatograms from the same sample to calibrate the uncertainty in the measurements each time a new batch of analyses is performed. The uncertainty is used to fix a threshold which distinguishes significant from insignificant distances ("distance threshold"). The distance threshold is about the same as the error for replicate analysis. Hereby, the method builds the distribution histogram of the differences between two chromatograms from the same sample. The shape of the histogram is used to precisely determine the uncertainty in the measurements. This uncertainty is used to obtain the expected profile of the distribution that the differences between two chromatograms should have if they were from the same sample (Nouvelle & Coutrot, 2010).

For GC fingerprinting purposes, the peak quality depends on two main quantities:

- Retention time,
- Peak height ratios

Peak height ratios are generally used for GC fingerprinting. In contrast to peak heights, the peak high ratios avoid the uncertainties dealing with discrepancies between the lightest and the heaviest compounds in the samples. The origin of such discrepancies can be linked to sampling procedure issues, evaporation of the lightest hydrocarbons or also blocking of the heaviest hydrocarbons into the column or the injection device. For all GC runs the retention times or the peak height ratios need to be checked on the tolerance deviations.

The peak height ratios are calculated from the indexation. In practice, several hundreds of peak height ratios are used. A maximal retention time difference between two peaks implied in a ratio is settled to 25 Kovats indices. A Kováts indice is a retention time measurement relative to two consecutive n-paraffins (Kováts, 1958):

$$KovatsID(i) = 100 \cdot \left(nC_{i-1} + \frac{\log_{10}(t_i) - \log_{10}(t_{C_{i-1}})}{\log_{10}(t_{C_i}) - \log_{10}(t_{C_{i-1}})} \right) \tag{1}$$

where:

- KovátsID(i): Kovats index of the compound,
- nC_{i-1}: number of carbon on the n-paraffin located just before the compound,
- t_i: retention time of the compound,
- $t_{C_{i-1}}$: retention time of the n-paraffin located just before the compound,
- t_{C_i} : retention time of the n-paraffin located just after the compound.

The aim of the indexation tool is to compare several chromatograms with a reference chromatogram. The peaks of the reference chromatogram are searched in the other chromatograms on the basis of a topological analysis. Figure 4.1 shows as example the chromatograms of two similar oils in the range of nC_{11} and nC_{12}. Even though both chromatograms are similar, they differ in certain peak height ratios, which clearly differentiate both chromatograms belonging to different oil samples.

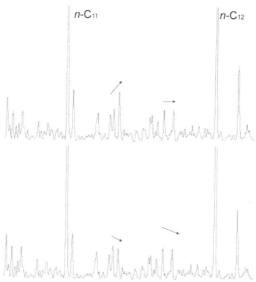

Fig. 4.1. Chromatographic comparison of 2 oil samples based on nC_{11}-nC_{12} inter-paraffin peak height ratios. Arrows indicate differences in neighboring peak height ratios, which indicates compositional differences between both crude oil samples.

The final outcome for a set of oil samples in terms of their similarity is commonly presented graphically as star diagram or cluster analysis. As example, Figure 4.2 shows the star diagram that compares 11 oils from a single well based on 37 n-$C_7 - n$-C_{20} range peak height ratios.

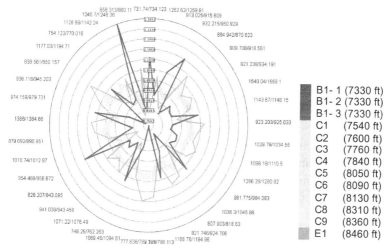

Fig. 4.2. Star diagram comparison of 11 oil samples from a single well based on 37 nC_7-nC_{20} range peak ratios. Each axis on the star shows the values for a different ratio of a pair of GC peaks. Peak labels are in Kovats Indices where peaks eluting from the GC between C_7 and C_8 are labelled in the 700's, those eluting between C_8 and C_9 are in the 800's.

Reduction of the number of variables was performed by using the Reduction by Inertia Constraint (RIC) method, with optimization by Inter/Intra-class maximization. The quality ω of each ratio is proportional to the standard deviation between the pre-groups, and inversely proportional to the standard deviation for analysis of samples within the same pre-group (Nouvelle, 2010). Thus, for a given ratio, the quality ω is the best when:

- The pre-groups are well separated.
- The analyses belonging to the same groups are close.

$$\omega_j = \frac{\sigma_j^{INTER}}{\sigma_j^{INTRA}} \qquad (2)$$

with σ_j^{INTER} : standard deviation calculated for inter-class maximization

        σ_j^{IINTRA} : standard deviation calculated for intra-class maximization

Triplicate analyses of the oil sample B are included to determine the distance threshold for the dataset. The replicate analysis of sample B is then grouped to determine the quality of the different variables attached to the dataset. Comparison of the star patterns indicates that the oil sample B (red; three replicate analyses) is significantly different than the remaining oils marked as oils C (blue) and oil E (yellow).

Another graphical evaluation method for the similarity of GC chromatograms is the hierarchical cluster analysis. Hierarchical clustering is general mathematical approaches, in

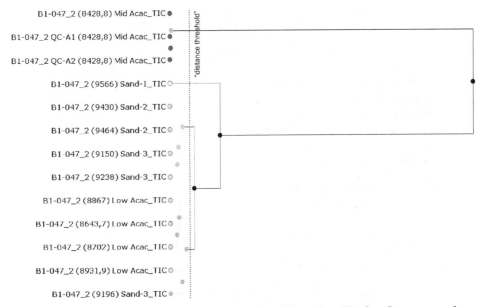

Fig. 4.3. Cluster analysis of the 11 oils from a single well based on 37 nC_7-nC_{20} range peak height ratios.

which the oils are grouped together that are closer to one another based on the distribution analysis using GC chromatograms. A key component of the analysis is repeated calculation of distance measures between data, and between clusters once oils begin to be grouped into clusters. The outcome is represented graphically as a dendrogram (Figure 4.3). For the same samples set of 11 oils in Figure 4.2, the dendogram is shown in Figure 4.3. The colouring of the groups on this diagram allows one to more easily discern the groups visually. Instrumental error is low as indicated by the tie line connection between the three replicate analyses of oil sample B.

5. Conclusions

Since its introduction, gas chromatography (GC) has been widely used as an imnportant method in the analysis of petroleum hydrocarbon which have complex compositions. New techniques have extended the applications of GC in the petroleum composition analysis. In this chapter, several such techniques have been briefly reviewed, including high temperature GC (HTGC), two-dimensional GC (2D-GC, or GC×GC) and GC fingerprinting. Although some of these techniques, such as 2D-GC and GC fingerprinting, are still very young, it is expected that, with the advance of the research work, they will used more routinely to give more precise composition in a broader range, and to give more important geochemical information of the petroleum fluids in the near future.

6. References

Adahchour, M., Beens, J & Brinkman, U. A. T. (2008). Recent developments in the application of comprehensive two-dimensional gas chromatography. *Journal of Chromatography. A*, 1186(1-2), pp. 67-108, ISSN 0021-9673

Adahchour, M., Beens, J., Vreuls, R. J. J. & Brinkman, U. A. T. (2006). Recent developments in comprehensive two-dimensional gas chromatography (GC × GC). II. Modulation and detection. *Trends in Analytical Chemistry*, 25(6), pp. 540-553, ISSN 0165-9936

Aguiar, A., Aguiar, H. G. M., Azevedo, D. A. & Aquino Neto, F. R. (2011). Identification of Methylhopane and Methylmoretane Series in Ceara Basin oils, Brazil, Using Comprehensive Two-Dimensional Gas Chromatography Coupled to Time-of-Flight Mass Spectrometry. *Energy and Fuels*, 25(3), pp. 1060-1065, ISSN 1520-5029

Asmussen, C. & Stan, H. (1998). Determination of non-ionic surfactants of the alcohol polyethoxylate type by means of high temperature gas chromatography and atomic emission detection. *Journal of High Resolution Chromatography*, 21(11), pp. 597-604, ISSN 1521-4168

Betancourt, S. S., Ventura, G. T., Pomerantz, A. E., Viloria, O., Dubost, F. X., Zuo, J., Monson, G., Bustamante, D., Purcell, J. M., Nelson, R.K., Rodgers, R. P., Reddy, C. M., Marshall, A. & Mullins O. C.(2009). Nanoaggregates of asphaltenes in a reservoir crude oil and reservoir connectivity. *Energy and Fuels*, 23(3), pp. 1178-1188, ISSN 1520-5029

Boczkaj, G., Przyjazny, A. & Kamiński, M. (2011). A new procedure for the determination of distillation temperature distribution of high-boiling petroleum products and fractions. *Analytical and Bioanalytical Chemistry*, 399(9), pp. 3253-3260, ISSN 1618-2650

Buenrostro-Gonzalez, E., Andersen, S. I., Garcia-Martinez, J. A. & Lira-Galeana, C. (2002). Solubility/molecular structure relationships of asphaltenes in polar and nonpolar media. *Energy and Fuels*, *16*(3), pp. 732-741, ISSN 1520-5029

Coto, B., Coutinho, J. A. P., Martos, C., Robustillo, M. D., Espada, J. J. & Peña, J. L. (2011). Assessment and improvement of n-Paraffin distribution obtained by HTGC to predict accurately crude oil cold properties. *Energy and Fuels*, *25*(3), pp. 1153-1160, ISSN 1520-5029

Damasceno, L. M. P., Cardoso, J. N. & Coelho, R. B. (1992). High temperature gas chromatography on narrow bore capillary columns. *Journal of High Resolution Chromatography*, *15*(4), pp. 256-259, ISSN 1521-4168

Davarnejad, R., Kassim, K. M., Zainal, A. & Suhairi. A. S. (2007). Phase equilibrium study in supercritical fluid extraction of ethanol to octane mixture using CO_2. *Asean Journal of Chemical Engineering*, *7*(2), pp. 127-136, 1655-4418

Davarnejad, R., Kassim, K. M., Zainal, A. & Suhairi. A. S. (2008). Mutual solubility study for 94.2:5.8 of ethanol to octane with supercritical carbon dioxide solvent. *Journal of Chinese Institute of Chemical Engineers*, *39*(4), pp. 343-352, ISSN 0368-1653

Dulaurent, A., Dahan, L., Thiébaut, D., Bertoncini, F. & Espinat, D. (2007). Extended simulated distillation by capillary supercritical fluid chromatography. *Oil and Gas Science and Technology*, *62*(1), pp. 33-42, ISSN 1294-4475

Dutriez, T., Borras, J., Courtiade, M., Thiébaut, Didier, Dulot, H., Bertoncini, F. & Hennion, M. C. (2011). Challenge in the speciation of nitrogen-containing compounds in heavy petroleum fractions by high temperature comprehensive two-dimensional gas chromatography. *Journal of Chromatography. A*, *1218*(21), pp. 3190-3199, ISSN 0021-9673

Dutriez, T., Courtiade, M., Thiébaut, Didier, Dulot, H., Bertoncini, Fabrice, Vial, J. & Hennion, M. C. (2009). High-temperature two-dimensional gas chromatography of hydrocarbons up to nC60 for analysis of vacuum gas oils. *Journal of Chromatography A*, *1216*(14), pp. 2905-2912, ISSN 0021-9673

Dutriez, T., Courtiade, M., Thiébaut, D., Dulot, H., Borras, J., Bertoncini, F. & Hennion, M. C. (2010). Advances in quantitative analysis of heavy petroleum fractions by liquid chromatography–high-temperature comprehensive two-dimensional gas chromatography: breakthrough for conversion processes. *Energy and Fuels*, *24*(8), pp. 4430-4438, ISSN 1520-5029

Glindemann, D., Ilgen, G., Herrmann, R. & Gollan, T. (2002). Advanced GC/ICP-MS design for high-boiling analyte speciation and large volume solvent injection. *Journal of Analytical Atomic Spectrometry*, *17*(10), pp. 1386-1389, ISSN 1364-5544

Gupta, A. K. & Severin, D. (1997). Characterization of petroleum waxes by high temperature gas chromatography - correlation with physical properties. *Petroleum Science and Technology*, *15*(9), pp. 943-957, ISSN 1532-2459

Heath, D. J., Lewis, C. A. & Rowland, S. J. (1997). The use of high temperature gas chromatography to study the biodegradation of high molecular weight hydrocarbons. *Organic Geochemistry*, *26*(11-12), pp. 769-785, ISSN 0146-6380

Hsieh, M. & Philp, R. P. (2001). Ubiquitous occurrence of high molecular weight hydrocarbons in crude oils. *Organic Geochemistry*, *32*(8), pp. 955-966, ISSN 0146-6380

Hsieh, M., Philp, R. P. & del Rio, J. C. (2000). Characterization of high molecular weight biomarkers in crude oils. *Organic Geochemistry*, *31*(12), pp. 1581-1588, ISSN 0146-6380

Huang, H., Larter, S. R. & Love, G. D. (2003). Analysis of wax hydrocarbons in petroleum source rocks from the Damintun depression, eastern China, using high temperature gas chromatography. *Organic Geochemistry*, *34*(12), pp. 1673-1687, ISSN 0146-6380

Juyal, P., McKenna, A. M., Yen, A., Rodgers, R. P., Reddy, C. M., Nelson, R. K., Andrews, A. B., Atolia, E., Allenson, S. J., Mullins, O. C. & Marshall, A. G (2011). Analysis and identification of biomarkers and origin of color in a bright blue crude oil. *Energy and Fuels*, *25*(1), pp. 172-182, ISSN 1520-5029

Kaal, E. & Janssen, H. G. (2008). Extending the molecular application range of gas chromatography. *Journal of Chromatography A*, *1184*(1-2), pp. 43-60, ISSN 0021-9673

Kelland, M. A. (2009). *Production Chemicals for the Oil and Gas Industry*. CRC Press, ISBN 978-142-0092-90-5

Kharrat, A. M. (2009). Characterization of Canadian heavy oils using sequential extraction approach. *Energy and Fuels*, *23*(2), pp. 828-834, ISSN 1520-5029

Kovát, E. (1958). Gas-chromatographische Charakterisierung organischer verbindungen. Teil 1: Retentionsindices aliphatischer Halogenide, Alkohole, Aldehyde und Ketone. *Helvetica Chimica Acta*, *41*(7), pp. 1915–1932, ISSN 1522-2675

Li, M., Zhang, S., Jiang, C., Zhu, G., Fowler, M., Achal, S., Milovic, M., Robinson, R. & Larter, S. (2008). Two-dimensional gas chromatograms as fingerprints of sour gas-associated oils. *Organic Geochemistry*, *39*(8), pp. 1144-1149, ISSN 0146-6380

van Lieshout, M., Janssen, H., Cramers, C. A., Hetem, M. J. & Schalk, H. (1996). Characterization of polymers by multi-step thermal desorption/programmed pyrolysis gas chromatography using a high temperature PTV injector. *Journal of High Resolution Chromatography*, *19*(4), pp. 193-199, ISSN 1521-4168

Liu, Z. & Phillips, J. B. (1991). Comprehensive two-dimensional gas chromatography using an on-column thermal modulator interface. *Journal of Chromatographic Science*, *29*(6), pp. 227-231, ISSN 0021-9665

Marriott, P., Dunn, M., Shellie, R. & Morrison, P. (2003). Targeted multidimensional gas chromatography using microswitching and cryogenic modulation. *Analytical Chemistry*, *75*(20), pp. 5532-5538, ISSN 0003-2700

Mayer, B. X., Rauter, W., Kählig, H. & Zöllner, P. (2003). A trifluoropropyl-containing silphenylene–siloxane terpolymer for high temperature gas chromatography. *Journal of Separation Science*, *26*(15-16), pp. 1436-1442, ISSN 1615-9314

Nielsen, N.-P. V., Carstensen, J. M. & Smedsgaard, J. (1998). Aligning of single and multiple wavelength chromatographic profiles for chemometric data analysis using correlation optimised warping. *Journal of Chromatography A*, *805*(1-2), pp. 17-35, ISSN 0021-9673

Nouvelle, X. & Coutrot, D. (2010). The Malcom distribution analysis method: A consistent guideline for assessing reservoir compartmentalisation from GC fingerprinting. *Organic Geochemistry*, *41*(9), pp. 981-985, ISSN 0146-6380

Phillips, J. B. & Beens, J. (1999). Comprehensive two-dimensional gas chromatography: a hyphenated method with strong coupling between the two dimensions. *Journal of Chromatography A*, *856*(1-2), pp. 331-347, ISSN 0021-9673

Philp, R. P. (1994). High temperature gas chromatography for the analysis of fossil fuels: A review. *Journal of High Resolution Chromatography, 17*(6), pp. 398-406, ISSN 1521-4168

Philp, R. P., Hsieh, M. & Tahira, F. (2004). An overview of developments related to the characterization and significance of high molecular weight paraffins/hydrocarbons (>C40) in crude oils. *Geological Society, London, Special Publications, 237*(1), pp. 37 -51, ISSN 0305-8719

Pursch, M., Sun, K., Winniford, B., Cortes, H., Weber, A., McCabe, T. & Luong, J. (2002). Modulation techniques and applications in comprehensive two-dimensional gas chromatography (GC×GC). *Analytical and Bioanalytical Chemistry, 373*(6), pp. 356-367, ISSN 1618-2650

del Rio, J. C. & Philp, R. P. (1992). High-molecular-weight hydrocarbons: A new frontier in organic geochemistry. *Trends in Analytical Chemistry, 11*(5), pp. 194-199, ISSN 0165-9936

Roehner, R. M., Fletcher, J. V., Hanson, F. V. & Dahdah, N. F. (2002). Comparative compositional study of crude oil solids from the Trans Alaska Pipeline System using high-temperature gas chromatography. *Energy and Fuels, 16*(1), pp. 211-217, ISSN 1520-5029

Subramanian, M., Deo, M. D. & Hanson, F. V. (1996). Compositional analysis of bitumen and bitumen-derived products. *Journal of Chromatographic Science, 34*(1), pp. 20-26, ISSN 0021-9665

Takayama, Y., Takeichi, T., Kawai, S. & Morikawa, M. (1990). Behaviours of siloxane polymers containing phenyl or silarylene as stationary phases for high-temperature gas chromatography. *Journal of Chromatography A, 514*, pp. 259-272, ISSN 0021-9673

Thanh, N. X., Hsieh, M. & Philp, R. P. (1999). Waxes and asphaltenes in crude oils. *Organic Geochemistry, 30*(2-3), pp. 119-132, ISSN 0146-6380

Tomasi, G., van den Berg, F. & Andersson, C. (2004). Correlation optimized warping and dynamic time warping as preprocessing methods for chromatographic data. *Journal of Chemometrics, 18*(5), pp. 231-241, ISSN 0886-9383

Vendeuvre, C., Ruiz-Guerrero, R., Bertoncini, F., Duval, L. & Thiébaut, D. (2007). Two-dimensional gas chromatography for detailed characterisation of petroleum products. *Oil & Gas Science and Technology - Revue de l'IFP, 62*(1), pp. 43-55, ISSN 1294-4475

Vendeuvre, C., Ruiz-Guerrero, R., Bertoncini, F., Duval, L., Thiébaut, D. & Hennion, M. C. (2005). Characterisation of middle-distillates by comprehensive two-dimensional gas chromatography (GC × GC): A powerful alternative for performing various standard analysis of middle-distillates. *Journal of Chromatography A, 1086*(1-2), pp. 21-28, ISSN 0021-9673

Wang, F. C.-Y., & Walters, C. C. (2007). Pyrolysis comprehensive two-dimensional gas chromatography study of petroleum source rock. *Analytical Chemistry, 79*(15), pp. 5642-5650, ISSN 0165-9936

Wang, Y. W., Chen, Q., Norwood, D. L. & McCaffrey, J. (2010). Recent development in the application of comprehensive two-dimensional gas chromatograph. *Journal of Liquid Chromatography & Related Technologies, 33*(9-12), pp. 1082-1115, ISSN 1082-6076

Woods, J. R., Kung, J., Kingston, D., Kotlyar, L., Sparks, B. & McCracken, T. (2008). Canadian crudes: A comparative study of SARA fractions from a modified HPLC separation technique. *Oil and Gas Science and Technology*, *63*(1 SPEC. ISS.), pp. 151-163, ISSN 1294-4475

Zrostlíková, J., Hajslová, J. & Cajka, T. (2003). Evaluation of two-dimensional gas chromatography-time-of-flight mass spectrometry for the determination of multiple pesticide residues in fruit. *Journal of Chromatography A*, *1019*(1-2), pp. 173-186, ISSN 0021-9673

Zuo, J. Y. & Zhang, D. (2000) Plus fraction characterization and PVT data regression for reservoir fluids near critical conditions. Paper SPE 64520 presented at the SPE Asia Pacific Oil and Gas Conference and Exhibition, Brisbane, Australia, 16–18, October, 2000.

Zuo, J. Y. & Zhang, D. (2008). Wax formation from synthetic oil systems and reservoir fluids. *Energy and Fuels*, *22*(4), pp. 2390-2395, ISSN 1520-5029

Pyrolysis-Gas Chromatography/Mass Spectrometry of Polymeric Materials

Peter Kusch
Bonn-Rhine-Sieg University of Applied Sciences,
Department of Applied Natural Sciences, Rheinbach,
Germany

1. Introduction

Structural analysis and the study of degradation properties are important in order to understand and improve performance characteristics of synthetic polymers and copolymers in many industrial applications. Polymers/copolymers cannot be analyzed in their normal state by traditional gas chromatography (GC) because of high molecular weight and lack of volatility. However, by heating these macromolecules to temperatures above 500 °C, they are pyrolyzed into many individual fragmentation substances, which can be then separated chromatographically and identified by mass spectrometry.

Pyrolysis technique hyphenated to gas chromatography/mass spectrometry (GC/MS) has extended the range of possible tools for characterization of synthetic polymers/copolymers. Under controlled conditions at elevated temperature (500 – 1400 °C) in the presence of an inert gas, reproducible decomposition products characteristic of the original polymer/copolymer sample are formed. The pyrolysis products are chromatographically separated by using a fused silica capillary column and subsequently identified by interpretation of the obtained mass spectra or by using mass spectra libraries (*e.g.* NIST, Wiley, etc.). Pyrolysis methods eliminate the need for pre-treatment by performing analyses directly on the solid polymer/copolymer sample.

Most of the thermal degradation results from free radical reactions initiated by bond breaking and depends on the relative strengths of the bonds that hold the molecules together. A large molecule will break apart and rearrange in a characteristic way (Moldoveanu, 2005; Wampler, 2007; Sobeih et al., 2008). If the energy transfer to the sample is controlled by temperature, heating- rate and time, the fragmentation pattern is reproducible and characteristic for the original polymer. Another sample of the same composition heated at the same rate to the same temperature for the same period of time will produce the same decomposition products. Therefore the essential requirements of the apparatus in analytical pyrolysis are reproducibility of the final pyrolysis temperature, rapid temperature rise and accurate temperature control. Depending upon the heating mechanism, pyrolysis systems have been classified into two groups: the continuous-mode pyrolyzer (furnace pyrolyzer) and pulse-mode pyrolyzer (flash pyrolyzer) such as the heated filament, Curie-point and laser pyrolyzer. The pyrolysis unit is directly connected to the injector port of a gas chromatograph. A flow of an inert carrier gas, such as helium, flushes the pyrolyzates into the fused silica capillary column. The detection technique of the separated compounds is typically mass spectrometry but other GC detectors have been also used depending on the intentions of the analysis (Sobeih et al., 2008). The currently commercially available pyrolysis equipment was described in detail in previous work of the author (Kusch et al., 2005).

The applications of analytical pyrolysis–gas chromatography/mass spectrometry range from research and development of new materials, quality control, characterization and competitor product evaluation, medicine, biology and biotechnology, geology, airspace, environmental analysis to forensic purposes or conservation and restoration of cultural heritage. These applications cover analysis and identification of polymers/copolymers and additives in components of automobiles, tires, packaging materials, textile fibers, coatings, half-finished products for electronics, paints or varnishes, lacquers, leather, paper or wood products, food, pharmaceuticals, surfactants and fragrances.

In earlier publications of the author (Kusch, 1996; Kusch et al., 2005), the analysis and identification of degradation products of commercially available synthetic polymers and copolymers by using analytical pyrolysis hyphenated to gas chromatography/FID and gas chromatography/mass spectrometry have been presented. In this chapter, examples of application of this analytical technique for identification of different polymeric materials are demonstrated.

2. Experimental

2.1 Instrumentation

Approximately 100 – 200 µg of solid sample were cut out with scalpel and inserted without any further preparation into the bore of the pyrolysis solids-injector and then placed with the plunger on the quartz wool of the quartz tube of the furnace pyrolyzer *Pyrojector II™* (S.G.E., Melbourne, Australia) (Figs. 1–2). The pyrolyzer was operated at a constant temperature of 550 °C, 600 °C or 700 °C. The pressure of helium carrier gas at the inlet to the furnace was 95 kPa. The pyrolyzer was connected to a *7890A* gas chromatograph with a series *5975C* quadrupole mass spectrometer (Agilent Technologies Inc., Santa Clara, CA, U.S.A.) operated in electron impact ionization (EI) mode. Two fused silica capillary columns (1) 60 m long, 0.25 mm I. D. and (2) 59 m long, 0.25 mm I. D. with *DB-5ms* stationary phase, film thickness 0.25 µm were used. Helium, grade 5.0 (Westfalen AG, Münster, Germany) was used as a carrier gas. The gas chromatographic (GC) conditions were as follow:

-    (1) programmed temperature of the capillary column from 60 °C (1 min hold) at 7 °C min^{-1} to 280 °C (hold to the end of analysis) and programmed pressure of helium from 122.2 kPa (1 min hold) at 7 kPa/min to 212.9 kPa (hold to the end of analysis),
-    (2) programmed temperature of the capillary column from 75 °C (1 min hold) at 7 °C min^{-1} to 280 °C (hold to the end of analysis) and programmed pressure of helium from 122.2 kPa (1 min hold) at 7 kPa/min to 212.9 kPa (hold to the end of analysis),
-    (3) programmed temperature of the capillary column from 60 °C (1 min hold) at 7 °C min^{-1} to 280 °C (hold to the end of analysis) and constant helium flow of 1 cm^3 min^{-1} during the whole analysis.

The temperature of the split/splitless injector was 250 °C and the split ratio was 20 : 1. The transfer line temperature was 280 °C. The EI ion source temperature was kept at 230 °C. The ionization occurred with a kinetic energy of the impacting electrons of 70 eV. The quadrupole temperature was 150 °C. Mass spectra and reconstructed chromatograms (total ion current [TIC]) were obtained by automatic scanning in the mass range m/z 35 – 750 u. GC/MS data were processed with the *ChemStation* software (Agilent Technologies) and the *NIST 05* mass spectra library (Agilent Technologies).

Fig. 1. Pyrolysis-GC/MS system used in this work equipped with a furnace pyrolyzer *Pyrojector II*™ (S.G.E.), control module (S.G.E.), a *7890A* gas chromatograph and a series *5975C* quadrupole mass spectrometer (Agilent Technologies).

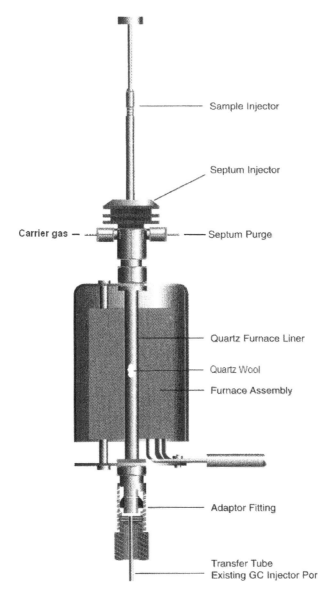

Fig. 2. Schematic view of the furnace pyrolyzer *Pyrojector II*™ (S.G.E., Melbourne, Australia).

2.2 Samples

Samples of an unknown industrial plastic, a valve rubber, a car tire rubber, membranes of hydraulic cylinders from the automotive industry, an O-ring seal, dental filling material and recycled polyethylene were used in the investigation.

3. Results and discussion

3.1 Analytical pyrolysis of synthetic organic polymers/copolymers

Pyrolysis–gas chromatography/mass spectrometry (Py–GC/MS) is used to characterize the structure of synthetic organic polymers and copolymers, polymer blends, biopolymers and natural resins. The traditional physical methods may only be applied to the analysis of technical organic polymers. Several chemical analysis techniques like UV-, FTIR- and NMR-spectroscopy, thermogravimetric analysis (TGA), size–exclusion chromatography (SEC, GPC), headspace–GC (HS–GC) or recently solid–phase microextraction (SPME) with GC/MS (Kusch & Knupp, 2002, 2004; Hakkarainen, 2008) have been established during the last four decades for characterization of macromolecules. These non-destructive methods offer information about functional groups, structural elements, thermal stability, molecular weight, and volatile components.

Pyrolysis–GC/MS is a destructive analytical technique. The most frequent use of this technique is the analysis of polymers and copolymers. Typical fields of interest and application (Hummel & Fischer, 1974; Hummel & Scholl, 1988; Hallensleben & Wurm, 1989; Moldoveanu, 2005; Kusch et al., 2005; Wampler, 2007) are:

- polymer identification by comparison of pyrograms and mass spectra with known references,
- qualitative analysis and structural characterization of copolymers, sequence statistics of copolymers, differentiation between statistical and block polymers,
- determination of the (micro) structure of polymers (degree of branching and cross-linking, compositional analysis of copolymers and blends, co-monomer ratios, sequence distributions, analysis of end-groups),
- determination of the polymers steric structure (stereoregularity, tacticity, steric block length, and chemical inversions),
- investigation of thermal stability, degradation kinetics and oxidative thermal decomposition of polymers and copolymers,
- determination of monomers, volatile organic compounds (VOC), solvents and additives in polymers,
- kinetic studies,
- quality control,
- quantification.

Starting at the University of Cologne (Germany) in the 1960s, Hummel and co-workers studied the process of thermal decomposition of polymers in detail using Py–GC/MS and Py–GC/FTIR (Hummel & Fischer, 1974; Hummel & Scholl, 1988). In general, decomposition proceeds through radical formation, which, due to the high reactivity of radicals, initiates numerous consecutive and parallel reactions. The authors summarized the main pathways of polymer decomposition in four categories:

1. retropolymerization from the end of the polymer-chain, predominantly forming monomers [e.g. poly(methyl methacrylate) (PMMA), poly-a-methylstyrene],
2. statistical chain scission followed by:
   - retropolymerization from radical bearing chain ends (e.g. polyisobutylene, polystyrene),
   - radical transfer and disproportionation (e.g. polyethylene, isotactic polypropylene),

- stabilization of fragments by cyclization (*e.g.* polydimethylsiloxane),
3. splitting side-chain leaving groups (*e.g.* polyvinylchloride, polyacrylonitrile, polyacrylate)
4. intramolecular condensation reactions with loss of smaller molecules (*e.g.* phenol-epoxide resins).

However, this classification is restricted to homogeneous polymers. The situation is more complex in copolymers, depending on the applied monomers and their linking (Hummel & Fischer, 1974; Hummel & Scholl, 1988; Kusch & Knupp, 2007).

In the following, practical application of the analytical pyrolysis connected to GC/MS for the identification of chemical structure and additives of an unknown industrial plastic, rubber products, dental filling material or recycled polyethylene will be presented. The possibility of using this technique in the failure analysis will be also demonstrated.

3.1.1 Identification of an unknown plastic sample

A sample of an industrial plastic was pyrolyzed at 600 °C and 700 °C, respectively in order to identify its composition. The total ion chromatograms (pyrograms) obtained for both pyrolysis temperatures were similar. Figure 3 shows the Py–GC/MS chromatogram of the sample pyrolyzed at 600 °C. Based on the decomposition products (Table 1), the plastic was identified as flexible poly(vinyl chloride) (PVC) with di-(2-ethylhexyl)phthalate (DEHP) plasticizer. The main features of the pyrolysis of PVC are the formation of hydrochloride (HCl) and the formation of unsaturated aliphatic and aromatic hydrocarbons. Benzene is the major pyrolysis product of PVC. Other aromatic components like toluene and the PAHs (polycyclic aromatic hydrocarbons) are also present (Table 1). This is the result of the formation of double bonds by the elimination of HCl from the PVC macromolecules, followed by the breaking of the carbon chain with or without cyclization (Moldoveanu, 2005). The formation of aromatic compounds in PVC pyrolysis is schematically exemplified for benzene in the same monograph (Moldoveanu, 2005).

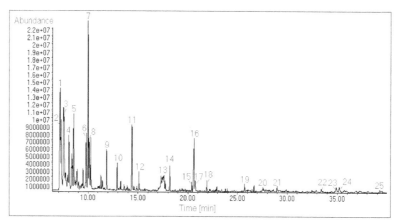

Fig. 3. Pyrolysis–GC/MS chromatogram of a plastic sample pyrolyzed at 600 °C, identified as flexible poly(vinyl chloride) (PVC). For peak identification, see Table 1. GC column 1, GC conditions 2.

The decomposition of the plasticizer di-(2-ethylhexyl)phthalate (DEHP) leads to the formation of 2-ethyl-1-hexene (RT = 9.98 min), 2-ethylhexanal (RT = 12.93 min), 2-ethyl-1-hexanol (RT = 14.40 min) and phthalic anhydride (RT = 20.63 min) (Table 1 and Fig. 3).

The obtained EI mass spectra of the compounds are shown in Fig. 4, while the chemical reaction of the thermal degradation of di-(2-ethylhexyl)phthalate is presented in Fig. 5. The reaction scheme is consistent with the previously published work (Bove & Dalven, 1984).

Peak	Retention time (RT) [min]	Pyrolysis products of polyvinyl chloride (PVC) with DEHP plasticizer at 600 °C
1	7.14	Hydrochloride/Propene
2	7.25	1-Butene
3	7.49	1,4-Pentadiene
4	8.05	2,4-Hexadiene
5	8.53	Benzene
6	9.81	Toluene
7	9.98	2-Ethyl-1-hexene
8	10.09	3-Methyl-3-heptene
9	11.89	Styrene
10	12.93	2-Ethylhexanal
11	14.40	2-Ethyl-1-hexanol
12	15.10	Indene
13	17.57	Benzoic acid
14	18.18	Naphthalene
15	20.38	2-Methylnaphthalene
16	20.63	Phthalic anhydride
17	20.73	1-Methylnaphthalene
18	21.91	Biphenyl
19	25.71	Fluorene
20	27.56	1,2-Diphenylethylene
21	28.97	Anthracene
22	33.45	Fluoranthene
23	34.89	Benzofluorene isomer
24	35.19	Benzofluorene isomer
25	39.35	Benz[a]anthracene

Table 1. Pyrolysis products of flexible polyvinyl chloride (PVC) with DEHP plasticizer at 600 °C. Peak numbers as in Fig. 3. GC column 1, GC conditions 2.

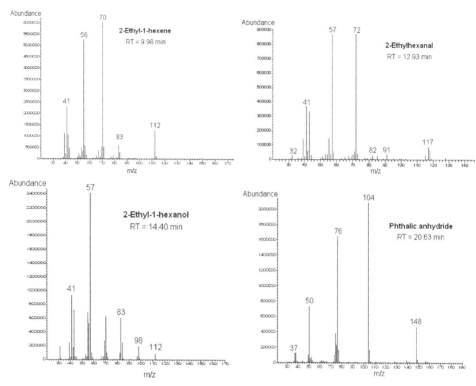

Fig. 4. EI ionization mass spectra at 70 eV of substances formed by the pyrolysis of the plasticizer di-(2-ethylhexyl)phthalate (DEHP) at 600 °C; RT = retention time of substance, as in Fig. 3 and in Table 1.

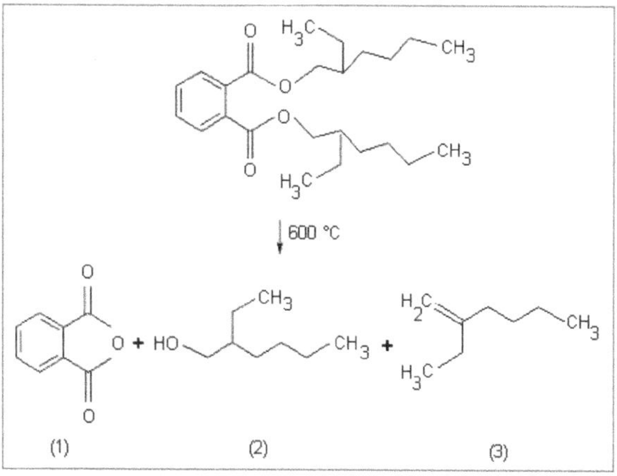

Fig. 5. Chemical reaction of the thermal decomposition of di-(2-ethylhexyl)phthalate (DEHP) at 600 °C leading to obtain (1) phthalic anhydride, (2) 2-ethyl-1-hexanol and (3) 2-ethyl-1-hexene.

3.1.2 Identification of rubber materials

The study of rubbers is the oldest application of analytical pyrolysis-GC (Kusch et al., 2005). Rubbers are frequently filled with opaque materials like carbon black, making them difficult for analysis by spectroscopy. Furthermore, cross-linking makes them insoluble and thus many of the traditional analytical tools for organic analysis are difficult or impossible to apply. Each rubber compound in a tire contains rubber polymers, sometimes one, but often a blend of two or more. The more commonly used polymers are natural rubber (NR, polyisoprene), synthetic polyisoprene (IR), polybutadiene (BR), and styrene-butadiene copolymers (SBR). SBR is widely used for tread compounds of a tire, generally in the tire treads of passenger cars, and NR in the relatively large-sized tires of buses and trucks etc. Inner tubes are usually based on butyl-rubber, a copolymer of isobutylene with a small proportion of isoprene.

Nitrile rubber (NBR) was invented at about the same time as SBR in the German program, in the end of the 1920s, to find substitutes for natural rubber (Graves, 2007). These rubbers are copolymers of acrylonitrile and 1,3-butadiene, containing 15 – 40% acrylonitrile. The major applications for this material are in areas requiring oil and solvent resistance. The largest market for nitrile rubber is in the automotive area because of its solvent and oil resistance. Major end uses are for hoses, fuel lines, O-rings, gaskets and seals. In blends with PVC and ABS, nitrile rubber acts as an impact modifier. Some nitrile rubber is sold in latex form for the production of grease-resistant tapes, gasketing material and abrasive papers. Latex also is used to produce solvent resistant gloves (Graves, 2007).

The criteria for assessing the quality of rubber materials are the polymer/copolymer composition and the additives. Commercial plastics and rubbers always contain a number of additives that are included to give particular physical and/or chemical properties. These additives include plasticizers, extender oils, carbon black, inorganic fillers, antioxidants, heat- and light stabilizers, tackifying resins, processing aids, cross-linking agents, accelerators, retarders, adhesives, pigments, smoke and flame retardants, and others (Hakkarainen, 2008).

The examined rubber materials in our laboratory were a valve rubber, a car tire rubber and an O-ring seal. On the one hand, based on the decomposition products at 700 °C (Fig. 6 and Table 2), the valve rubber was identified as a blend of polyisoprene (NR) and poly(styrene-co-butadiene) (SBR). On the other hand, the car tire rubber was determined as the mixture of polybutadiene (BR) and poly(styrene-co-butadiene) (SBR) (Kusch & Knupp, 2009).

Compounds, such as additives, are usually applied in low concentrations in plastic and rubbers. They are often recognizable as small irregularities within the characteristic pyrogram of the pyrolyzed material. By using the GC column 1 and the GC conditions 1, it was possible to identify the organic additives in both rubber samples. The peak of benzothiazole (Fig. 6, Peak 21, RT = 20.37 min) corresponds to the product of the thermal decomposition of benzothiazole-2-thiol (2-mercaptobenzothiazole, 2-MBT). In the rubber industry benzothiazole-2-thiol is used as vulcanisation accelerator and as antioxidant.

Furthermore, the antioxidant 2,6-bis-(1,1-dimethylethyl)-4-methylphenol (BHT) was detected in the pyrolyzate of the valve rubber (Fig. 6, Peak 29, RT = 25.50 min). The mass spectra obtained for both additives are shown in Fig. 7.

Peak	Retention time (RT) [min]	Pyrolysis products of the valve and the car tire rubbers at 700 °C
1	8.20	2-Butene
2	8.55	2-Methyl-1,3-butadiene
3	9.34	3-Methyl-2-pentene (isoprene)
4	9.75	5-Methyl-1,3-cyclopentadiene
5	10.08	Benzene
6	11.28	1-Methyl-1,4-cyclohexadiene
7	11.84	Toluene
8	13.51	Ethylbenzene
9	13.69	p-Xylene
10	14.08	Styrene
11	15.19	m-Ethyltoluene
12	15.52	a-Methylstyrene
13	15.81	1,2,4-Trimethylbenzene (pseudocumene)
14	16.28	1,2,3-Trimethylbenzene (hemimelitene)
15	16.78	Indene
16	17.43	o-Isopropenyltoluene
17	18.02	1,2,4,5-Tetramethylbenzene (durene)
18	18.75	3-Methylindene
19	18.86	2-Methylindene
20	19.58	Naphthalene
21	20.37	Benzothiazole
22	21.83	2-Methylnaphthalene
23	22.17	1-Methylnaphthalene
24	23.43	Biphenyl
25	24.06	Dimethylnaphthalene isomer
26	24.32	Dimethylnaphthalene isomer
27	24.43	Dimethylnaphthalene isomer
28	25.48	3-Methyl-1,1'-biphenyl
29	25.50	2,6-Bis-(1,1-dimethylethyl)-4-methylphenol (BHT)
30	26.52	1,6,7-Trimethylnaphthalene
31	27.75	Fluorene
32	29.88	1,2-Diphenylethylene (stilbene)
33	31.61	Anthracene
34	34.79	2-Phenylnaphthalene

Table 2. Pyrolysis products of the valve rubber and the car tire rubber materials at 700 °C. Peak numbers as in Fig. 6. GC column 1, GC conditions 1.

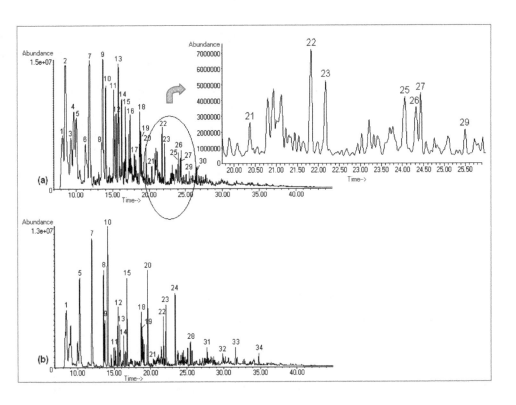

Fig. 6. Pyrolysis–GC/MS chromatograms (TIC) of the rubber materials at 700 °C;
(a) NR/SBR, (b) BR/SBR. Peak 21 = benzothiazole (RT = 20.37 min), peak 29 = BHT
(RT = 25.50 min). For other peak identification, see Table 2. GC column 1, GC conditions 1.

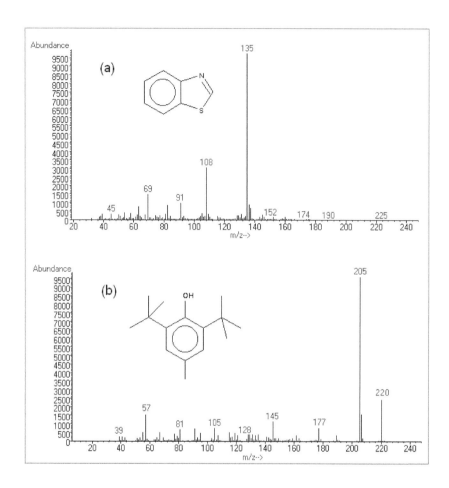

Fig. 7. EI mass spectra at 70 eV of the rubber additives pyrolyzed at 700 °C: (*a*) benzothiazole (thermal degradation product of benzothiazole-2-thiol), (*b*) 2,6-bis-(1,1-dimethylethyl)-4-methylphenol (BHT).

The pyrolysis–GC/MS chromatogram of the O-ring sample at 700 °C is shown in Fig. 8, while the identified degradation products are summarized in Table 3. Based on the decomposition products at 700 °C (Fig. 8 and Table 3), the O-ring rubber was identified as poly(acrylonitrile-*co*-butadiene) (NBR) with high content of the plasticizer di-(2-ethylhexyl)phthalate (DEHP) and with the vulcanization accelerator benzothiazole-2-thiol (2-mercaptobenzothiazole, 2-MBT). The analysis of pyrolyzate of poly(acrylonitrile-*co*-butadiene) indicated in Table 3 shows the presence of compounds generated from the acrylomitrile sequences (acrylonitrile, methacrylonitrile, benzonitrile), from the butadiene sequences (1,3-butadiene) and from the additives of the copolymer (2-ethyl-1-hexene, 2-ethyl-1-hexanol, phthalic anhydride and benzothiazol).

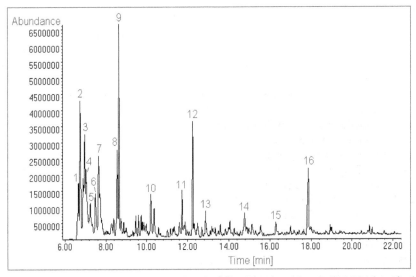

Fig. 8. Pyrolysis–GC/MS chromatogram (TIC) of the O-ring rubber at 700 °C identified as NBR with additives. For peak identification, see Table 3. GC column 2, GC conditions 2.

Peak	Retention time (RT) [min]	Pyrolysis products of the O-ring rubber at 700 °C
1	6.66	Propylene
2	6.74	1,3-Butadiene
3	6.97	Acrylonitrile
4	7.04	1,3-Cyclopentadiene
5	7.24	Methacrylonitrile
6	7.50	1,4-Cyclohexadiene
7	7.65	Benzene
8	8.56	Toluene
9	8.65	2-Ethyl-1-hexene
10	10.20	Styrene
11	11.73	Benzonitrile
12	12.24	2-Ethyl-1-hexanol
13	12.87	Indene
14	14.76	3-Methylindene
15	16.29	Benzothiazole
16	17.87	Phthalic anhydride

Table 3. Pyrolysis products of the O-ring rubber at 700 °C. Peak numbers as in Fig. 8. GC column 2, GC conditions 2.

3.1.3 Application of pyrolysis–GC/MS in the dentistry

A number of dental filling materials are presently available for tooth restorations. The four main groups of these materials, which dentists have used for about 35 years, are the conventional glass-ionomer cements, resin-based composites, resin-modified glass-ionomer cements and polyacid-modified resinous composities (Rogalewicz et al., 2006a). Light-curing glass-ionomer cements contain polyacrylic acid, chemically and/or photo-curing monomers (multifunctional methacrylates, like triethylene glycol dimethacrylate or 2-hydroxyethyl methacrylate), an ion-leaching glass and additives (initiators, inhibitors, stabilizers and others) (Rogalewicz et al., 2006a). Resin-modified glass-ionomer cements are now widely used in dentistry as direct filling materials, liners, bases, luting cements and fissure sealants (Rogalewicz et al., 2006b). These materials mainly consist of polymer matrix and glass-ionomer parts. The polymer matrix is based on a monomer system and different multifunctional methacrylates with additives (Rogalewicz et al., 2006b). Methacrylic monomers, like bisphenol A glycidyl methacrylate (Bis-GMA), urethane dimethacrylate (UDMA), triethylene glycol dimethacrylate (TEGDMA) and 2-hydroxyethyl methacrylate (HEMA) are the main components of resin-based dental filling materials. The presence of additives such as initiators, activators, inhibitors and plasticizers in uncured dental material mixture is necessary (Rogalewicz et al., 2006b).

Figure 9 shows the total ion current Py–GC/MS chromatogram of commercially light-curing dental filling material pyrolyzed at 550 °C. The pyrolysis products identified by using mass spectra library *NIST 05* are summarized in Table 4. The carbon dioxide (RT = 5.85 min) identified in pyrolyzate is formed from polyacrylic acid. The identified substances, like HEMA (RT = 13.65 min), EGDMA (RT = 19.48 min) and TEDMA (RT = 28.72 min) are known as standard composites of dental filling materials. Other compounds in Table 4, such as bisphenol A (RT = 33.10 min) or bisphenol A diglycidylether (RT = 42.42 min) are probably formed by thermal degradation of bisphenol A diglycidyl mono- or dimethacrylates. The presence of the additives, like antioxidant BHT [2,6-bis-(1,1-

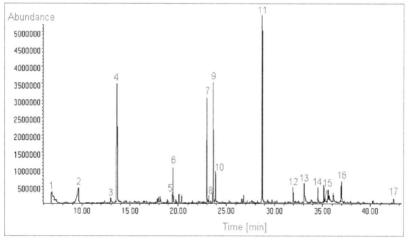

Fig. 9. Pyrolysis–GC/MS chromatogram of commercially light-curing dental filling material pyrolyzed at 550 °C. For peak identification, see Table 4. GC column 2, GC conditions 1.

dimethylethyl)-4-methylphenol, RT = 23.17 min] or the UV-absorber drometrizol (RT = 31.95 min) was also confirmed. The triphenylantimony (RT = 34.55 min) identified in pyrolyzate is used as catalyst in the UV-induced polymerisation.

Peak	Retention time (RT) [min]	Pyrolysis products of the dental filling material at 550 °C
1	5.85	Carbon dioxide
2	9.62	Methacrylic acid
3	12.96	Phenol
4	13.65	2-Hydroxyethyl methacrylate (HEMA)
5	19.40	4-Isopropenylphenol
6	19.48	Ethylene glycol dimethacrylate (EGDMA)
7	23.00	Not identified
8	23.17	2,6-Bis-(1,1-dimethylethyl)-4-methylphenol (BHT)
9	23.65	Not identified
10	23.89	Triethylene glycol (TEG)
11	28.72	Triethylene glycol dimethacrylate (TEDMA)
12	31.95	Drometrizol (Tinuvin-P, UV-absorber)
13	33.10	4,4´-Dihydroxy-2,2-diphenylpropane (Bisphenol A)
14	34.55	Triphenylantimony
15	35.25	Tetraethylene glycol dimethacrylate
16	36.98	4,4´-(1-Methylethylidene)-bis-[2,6-dimethylphenol]
17	42.42	Bisphenol A diglycidylether

Table 4. Pyrolysis products of commercially light-curing dental filling material at 550 °C. Peak numbers as in Fig. 9. GC column 2, GC conditions 1.

3.1.4 Environmental application of pyrolysis–GC/MS

Analytical pyrolysis has been used to study a variety of environmental samples including fossil fuel source rocks, natural resins or aquatic and terrestrial natural organic matter (White et al., 2004; Kusch et al., 2008a). Much work has been published on the analysis of soil (Tienpomt et al., 2001; Evans et al., 2003; Campo et al., 2011). Pyrolysis–GC/MS was also used in plastic recycling for analysis of products generated from the thermal and catalytic degradation of pure and waste polyolefins (Aguado et al., 2007).

Plastic recycling can be divided into mechanical (material) recycling, feedstock (chemical) recycling and energy recovery. In case of material recycling, the plastic waste is washed, ground and used in the plastic processing industry as a raw material. The chemical structure of the preserved material (re-granulate) is maintained. In case of chemical recycling, the residual polymers are converted catalytically into their monomers. The resulting monomers are then used in the manufacture of new plastics. In the energy recovery the plastic waste is used in cement- or in steel industry as an energy carrier.

The following example presents that pyrolysis–GC/MS allows the identification of contaminants in low-density polyethylene (LDPE) re-granulate from mechanical recycling process. Figure 10 shows the obtained Py–GC/MS chromatogram of the LDPE pellets at 700 °C. The pyrogram consists of serial triplet-peaks of straight-chain aliphatic C_3–C_{33} hydrocarbons, corresponding to α,ω-alkadienes, α-alkenes and n-alkanes, respectively, in the order of the increasing $n+1$ carbon number in the molecule. Such elution pattern is characteristic for the pyrolysis of polyethylene. For example, the triplet-peaks in Fig. 10 of C_{10} hydrocarbons correspond to 1,9-decadiene (RT = 14.01 min), 1-decene (RT = 14.16 min) and n-decane (RT = 14.33 min), respectively. The identification of the compounds was carried out by comparison of retention times and mass spectra of standard substances, study of the mass spectra, and comparison with data in the *NIST 05* mass spectra library. The substances identified in pyrolyzate, like 2,4-dimethyl-1-heptene (RT = 11.12 min) and styrene (RT = 12.39 min) (Fig. 10) indicate the contamination of the recycled polyethylene with polypropylene and polystyrene, respectively. The detected palmitic acid (RT = 31.27 min) and oleic acid (RT = 33.99 min) suggest the presence of plant residues in the investigated re-granular material.

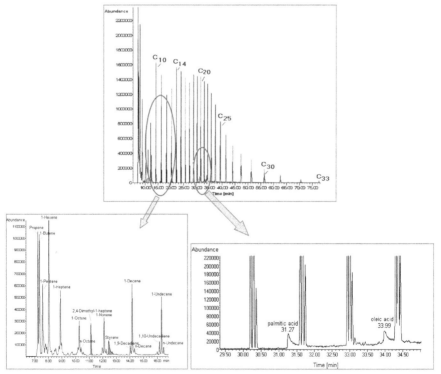

Fig. 10. Full view (top) and time windows between 6.2 – 16.8 min (bottom left) and 29.3 – 35.0 min (bottom right) of the Py–GC/MS chromatogram of recycled HDPE pyrolyzed at 700 °C. GC column 1, GC conditions 3.

3.1.5 Application of pyrolysis–GC/MS in the failure analysis

The increasing use of polymeric materials in the automotive industry requires sensitive and reliable methods for its analysis. For the failure analysis in motor vehicles there is often lack of information about the component itself, such as chemical composition, temperature resistance, possible contaminants or mechanical properties. The damage range is usually limited and not always homogeneous. There are often only small amounts of samples available to clarify the damage, which may be important for recognizing the cause of damage. Traditional analytical techniques used for characterization of polymers/copolymers, such as thermal analysis (TA) and Fourier transform infrared spectroscopy (FTIR) have limitations or are not sufficiently sensitive to demonstrate the change of the structure and the resulting dysfunction of used materials.

Previous work of the author (Kusch et al., 2008b) presents the application examples of Py–GC/MS in the failure analysis of various plastic or metal components and assemblies from the automotive industry. In the following case a glue residue on the surface of a rubber membrane from the hydraulic cylinder from the automotive industry was identified. Figure 11 shows the pyrolysis–GC/MS chromatograms at 700 °C, obtained from the new and operationally stressed rubber membrane coated with the glue residue, respectively.

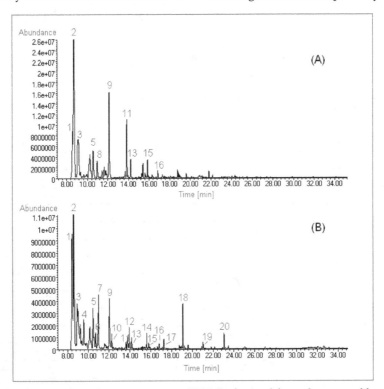

Fig. 11. Pyrolysis–GC/MS chromatograms at 700 °C, obtained from the new rubber membrane (A) and from the operationally stressed rubber membrane coated with glue residue (B). For peak identification, see Table 5. GC column 1, GC conditions 1.

The pyrolysis products identified by using mass spectra library *NIST 05* are summarized in Table 5. On the one hand, the pyrogram from the new rubber membrane obtained in Fig. 11 (A) consists of the characteristic fragments formed by the thermal degradation of poly(styrene-*co*-butadiene) (SBR). On the other hand, as can be seen from Fig. 11 (B) and from Table 5, the chemical structures of the identified substances of the operationally stressed rubber membrane coated with the glue residue build three groups of compounds: (1) *n*-alkenes and *n*-alkanes, (2) aromatics, and (3) ester. The presence of *n*-alkenes and heavy *n*-alkanes in pyrolyzate is characteristic for the pyrolysis of mineral oil (motor oil). The methyl methacrylate monomer (RT = 10.94 min) was produced by the retropolymerization of poly(methyl methacrylate) (PMMA) at 700 °C. Other peaks, like 1,3-butadiene (RT = 8.61 min), benzene (RT = 10.43 min), toluene (RT = 12.04 min), *p*-xylene (RT = 13.84 min), styrene (RT = 14.22 min), trimethylbenzene (RT = 15.87 min) and indene (RT = 16.85 min) are typical pyrolysis products of SBR. Based on the obtained results, it was possible to identify the glue coating from the surface of the rubber membrane from the automotive industry as a mixture of PMMA and mineral oil. The analytical results were then used for troubleshooting and remedial action of the technological process.

Peak	Retention time (RT) [min]	Pyrolysis products of the rubber membrane with glue coating at 700 °C	Pyrolyzed material
1	8.47	Propene	SBR, mineral oil
2	8.61	1-Butene/1,3-Butadiene	SBR, mineral oil
3	8.95	1-Pentene	Mineral oil
4	9.55	1-Hexene	Mineral oil
5	10.43	Benzene	SBR
6	10.72	1-Heptene	Mineral oil
7	10.94	Methyl methacrylate	PMMA
8	11.00	2,4-Dimethyl-1,3-pentadiene	SBR
9	12.04	Toluene	SBR
10	12.26	1-Octene	Mineral oil
11	13.84	*p*-Xylene	SBR
12	13.95	1-Nonene	Mineral oil
13	14.22	Styrene	SBR
14	15.56	1-Decene	Mineral oil
15	15.87	Trimethylbenzene	SBR
16	16.85	Indene	SBR
17	17.25	1-Undecene	Mineral oil
18	19.10	1-Dodecene	Mineral oil
19	21.04	1-Tridecene	Mineral oil
20	23.05	1-Tetradecene	Mineral oil
21	30.0 – 34.0	Heavy *n*-alkanes	Mineral oil

Table 5. Pyrolysis products of the rubber membrane with glue coating at 700 °C. Peak numbers as in Fig. 11. GC column 1, GC conditions 1.

4. Conclusions

Pyrolysis–GC/MS has proved as a valuable technique for the analysis and identification of organic polymeric materials in the plastic and rubber industry, dentistry, environmental protection and in the failure analysis. This technique allows the direct analysis of very small sample amounts without the need of time-consuming sample preparation.

5. Acknowledgements

The author thanks the *Kompetenzplattform "Polymere Materialien"* at the Bonn-Rhine-Sieg University of Applied Sciences in Rheinbach and the firm *Dr. Obst Technische Werkstoffe GmbH* in Rheinbach, Germany for samples and excellent cooperation.

I also thank my daughter Maria Kusch, M.A. for critical reading of the manuscript.

6. References

Aguado, J.; Serrano, D. P. & San Miguel, G. (2007). Analysis of Products Generated from the Thermal and Catalytic Degradation of Pure and Waste Polyolefins using Py–GC/MS. *Journal of Polymers and the Environment*, Vol. 15, No. 2, pp. 107-118

Bove, J. L. & Dalven, P. (1984). Pyrolysis of Phthalic Acid Esters: Their Fate. *The Science of the Total Environment*, Vol. 36, No. 1, pp. 313-318

Campo, J.; Nierop, K. G. J.; Cammeraat, E.; Andrei, V. & Rubio J. L. (2011). Application of Pyrolysis–Gas Chromatography/Mass Spectrometry to Study Changes in the Organic Matter of Macro- and Microaggregates of a Mediterranean Soil Uppon Heating. *Journal of Chromatography A*, Vol. 1218, No. 30, pp. 4817-4827

Evans, C. J.; Evershed, R. P.; Black, H. I. J. & Ineson, P. (2003). Compound-Specific Stable Isotope Analysis of Soil Mesofauna Using Thermally Assisted Hydrolysis and Methylation for Ecological Investigations. *Analytical Chemistry*, Vol. 75, No. 22, pp. 6056-6062

Graves, D. F. (2007). Rubber, In: *Kent and Riegels Handbook of Industrial Chemistry and Biotechnology, Part 1*, J. A. Kent (Ed.), pp. 689-718, Springer Science + Business Media, ISBN 978-0-387-27842-1, e-ISBN 978-0-387-27843-8, New York, U.S.A.

Hakkarainen, M. (2008). Solid–Phase Microextraction for Analysis of Polymer Degradation Products and Additives, In: *Chromatography for Sustainable Polymeric Materials*, A.-C. Albertsson & M. Hakkarainen (Eds.), pp. 23-50, Springer Verlag, ISBN 978-3-540-78762-4, e-ISBN 978-3-540-78763-1, Berlin, Heidelberg, Germany

Hallensleben, M. L. & Wurm, H. (1989). Polymeranalytik. *Nachrichten aus Chemie, Technik und Laboratorium*, Vol. 37, No. 6, pp. M1-M45

Hummel, D. O. & Fischer, H. (1974). *Polymer Spectroscopy*, D. O. Hummel (Ed.), Verlag Chemie, ISBN 3527254110, Weinheim, Germany

Hummel, D. O. & Scholl, F. (1988). *Atlas der Polymer- und Kunststoffanalyse*, D. O Hummel & F. Scholl (Eds.), Hanser Verlag, ISBN 3-446-12586-8, München, Germany

Kusch, P. (1996). Application of the Curie–Point Pyrolysis – High Resolution Gas Chromatography for Analysis of Synthetic Polymers and Copolymers. *Chemia Analityczna (Warsaw)*, Vol. 41, No. 2, pp. 241- 252

Kusch, P. & Knupp, G. (2002). Analysis of Residual Styrene Monomer and Other Volatile Organic Compounds in Expanded Polystyrene by Headspace Solid–Phase

Microextraction   Followed   by   Gas   Chromatography   and   Gas
        Chromatography/Mass Spectrometry. *Journal of Separation Science,* Vol. 25, No. 8,
        pp. 539-542

Kusch, P. & Knupp, G. (2004). Headspace–SPME–GC–MS Identification of Volatile Organic
        Compounds Released from Expanded Polystyrene. *Journal of Polymers and the
        Environment,* Vol. 12, No. 2, pp. 83-87

Kusch, P.; Knupp, G. & Morrisson, A. (2005). Analysis of Synthetic Polymers and
        Copolymers by Pyrolysis–Gas Chromatography/Mass Spectrometry, In: *Horizons
        in Polymer Research,* R. K. Bregg (Ed.), pp. 141-191, Nova Science Publishers, ISBN 1-
        59454-412-3, New York, U.S.A.

Kusch, P. & Knupp, G. (2007). Identifizierung von Verpackungskunststoffen mittels
        Pyrolyse und Headspace–Festphasenmikroextraktion in Kombination mit der
        Gaschromatographie/Massenspektrometrie. *LC·GC Ausgabe in deutscher Sprache,*
        June, pp. 28-34

Kusch, P.; Fink, W.; Schroeder-Obst, D. & Obst, V. (2008a). Identification of Polymeric
        Residues in Recycled Aluminium by Analytical Pyrolysis–Gas Chromatography-
        Mass Spectrometry. *ALUMINIUM International Journal,* Vol. 84, No. 4, pp. 76-79

Kusch, P.; Obst, V.; Schroeder-Obst, D.; Knupp, G. & Fink, W. (2008b). Einsatz der Pyrolyse
        –GC/MS zur Untersuchung von Polymeren Materialien in der Schadenanalyse in
        der Automobilindustrie. *LC·GC Ausgabe in deutscher Sprache,* July/August, pp. 5-11

Kusch, P. & Knupp, G. (2009). Identifizierung anhand der Zersetzungsprodukte. *Nachrichten
        aus der Chemie,* Vol. 57, No. 6, pp. 682-685

Moldoveanu, S. C. (2005). *Analytical Pyrolysis of Synthetic Organic Polymers,* Elsevier, ISBN 0-
        444-51292-6, Amsterdam, the Netherlands

Rogalewicz, R.; Voelkel, A. & Kownacki, I. (2006a). Application of HS–SPME in the
        Determination of Potentially Toxic Organic Compounds Emitted from Resin–Based
        Dental Materials. *Journal of Environmental Monitoring,* Vol. 8, No. 3, pp. 377-383

Rogalewicz, R.; Batko, K. & Voelkel A. (2006b). Identification of Organic Extractables from
        Commercial   Resin-Modified   Glass-Ionomers   using   HPLC–MS.   *Journal   of
        Environmental Monitoring,* Vol. 8, No. 7, pp. 750-758

Sobeih, K. L.; Baron, M. & Gonzales-Rodrigues, J. (2008). Recent Trends and Developments
        in Pyrolysis–Gas Chromatography. *Journal of Chromatography A,* Vol. 1186, No. 1-2,
        pp. 51-66

Tienpont, B; David, F.; Vanwalleghem, F. & Sandra, P. (2001). Pyrolysis–Capillary Gas
        Chromatography–Mass Spectrometry for the Determination of Polyvinyl Chloride
        Traces in Solid Environmental Samples. *Journal of Chromatography A,* Vol. 911, No. 2,
        pp. 235-247

Wampler, T. P. (Ed.) (2007). *Applied Pyrolysis Handbook,* second edition, CRC Press, ISBN 1-
        57444-641-X, Boca Raton, U.S.A.

White, D. M.; Garland, D. S.; Beyer, L. & Yoshikawa, K. (2004). Pyrolysis–GC/MS
        Fingerprinting of Environmental Samples. *Journal of Analytical and Applied Pyrolysis,*
        Vol. 71, No. 1, pp. 107-118

8

Determination of Organometallic Compounds Using Species Specific Isotope Dilution and GC-ICP-MS

Solomon Tesfalidet

Department of Chemistry, Umeå University
Sweden

1. Introduction

Isotope dilution is used in speciation analysis to determine the concentration of specific chemical forms of an element, offering the possibility for correcting eventual loses or alteration of the species that can occur during the analytical process. Gas Chromatography (GC) is known as the strongest separation method used for separation of volatile and thermo stable organic compounds. Among the detection methods that are used, inductively coupled plasma - mass spectrometry (ICP-MS) is marked as the detector of choice for organometallic compounds. The coupling of GC with ICP-MS is extended by making use of the capability of ICP-MS to detect several isotopes simultaneously. Isotope dilution (ID) which was originally used in organic mass spectrometry is now extensively used in speciation analysis of inorganic compounds. This requires that the analyte of interest has more than one stable isotope. Chemical forms of mono-isotopic elements such as Arsenic (m/z 75) can therefore not be determined using isotope dilution. Other elements such as lead (4 isotopes), tin (10 isotopes), mercury (7 isotopes) and chromium (4 isotopes) have all been determined using isotope dilution [1]. According to IUPAC (International Union of Pure and Allied Chemistry) definition speciation of an element is *"the distribution of that element amongst defined chemical species in a system"* and speciation analysis is *"analytical activities of identifying and/or measuring the quantities of one or more individual chemical species in a sample"*. Information about the total concentration of an element is usually not enough for a satisfactory characterization of a sample. Inorganic tin is for example harmless compared to organotin compounds (OTCs). The total concentration of tin, determined in a sample contaminated with the more toxic OTCs, does therefore not give meaningful data for risk assessment, making speciation analysis of the organotin compounds very essential.

In ID a known amount of an isotopically enriched element is added to a sample material. A simple question that one may raise is: what does this addition of an isotopically enriched element lead to? To answer this question Mikael Berglund referred to a work done by an enthymologist named C.H.N Jackson back in 1933[2]. Jackson used fly dilution to study the density of tsetse flies in a region in the village Ujiji at Tanganyka lake in Africa [3]. Jackson released a 'synthetic" Tsetse fly population, which consisted of marked Tsetse flies, into a natural Tsetse fly population. After complete mixing he caught a representative sample of Tsetse flies and counted the marked and unmarked flies. Knowing the total number of marked and released flies in the sample he could calculate the number of unmarked flies in the region. In a similar fashion we can imagine of counting the number of blue balls (?) in container A in figure 1. To this we add a known amount of identical, red colored, balls from container B. After a thorough mixing we take a portion from the mixture M, and count how many of each ablue nd red balls we have in this portion {P}.

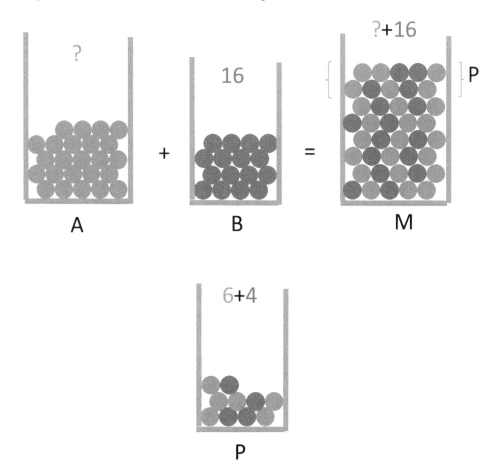

Fig. 1. A: container containing unknown number of blue balls; B: container containing known number of red balls; M: mixture of blue balls in A and red balls from B; P a portion of balls taken from the mixture M.

Using analogous equation to that used by Jackson back in 1933 we can calculate the number of blue bolls in M which also is equal to that we have in container A.

$$\frac{Number\ of\ red\ bolls\ in\ P}{Number\ of\ blue\ bolls\ in\ p} = \frac{Total\ number\ of\ red\ bolls\ in\ C}{Total\ number\ of\ blue\ bolls\ in\ C} \tag{1}$$

$$\frac{4}{6} = \frac{16}{Total\ number\ of\ blue\ bolls\ in\ C}$$

Total number of blue bolls in M = 24 = number of blue bolls in container A.

This quantification principle, exemplified by the ball addition, is in fact the same as that used in isotope dilution (ID) where isotopically marked analyte (called spike) is added to a sample, containing a natural analyte.

In ID a known amount of enriched isotope (called the "spike") is added to a sample. After equilibration of the spike isotope with the natural isotope in the sample the isotopic ratio is measured. Normally a mass spectrometer is used to measure the altered isotopic ratio(s). The measured isotope ratio of isotope A to isotope B, R_m, can be calculated using equation 1.

$$R_m = \frac{A_x C_x + A_s C_s W_s}{B_x C_x W_x + B_s C_s W_s}$$ (2)

where A_x and B_x are the atom fractions of isotopes A and B in the sample; A_s and B_s are the atom fractions of isotopes A and B in the spike; C_x and C_s, are the concentrations of the analyte in the sample and spike, respectively; and W_x and W_s are the weights of the sample and spike, respectively. The concentration of the analyte in the sample can then be calculated using equation 2.

$$C_x = \left(\frac{C_s W_s}{W_x}\right)\left(\frac{A_s - R_m B_s}{R_m B_x - A_x}\right)$$ (3)

It should however be noted that depending on the analytical method at hand, the stability of the species under investigation, and the need for addressing the resulting uncertainty in the analytical result the mathematical expression used for calculation of species concentration can look different and get more complicated [2, 4, 5]. In the forthcoming sections speciation analysis of organtin compounds will be discussed. Organotin compounds are one of the most investigated organometallic compounds, using isotope dilution GC-ICPMS, and will therefore fit perfect in this chapter for elaborating the analytical methodology and explain why isotope dilution is needed for speciation analysis.

2. Speciation of organotin compounds (OTCs)

Tin in its inorganic form is generally accepted as being non-toxic, but the toxicological pattern of organotin compounds is very complex. The biological effects of the substances depend on both the nature and the number of the organic groups bound to the Sn cation. Figure 2 shows the ionic form of the most extensively studied types of organotin compounds.

The ecotoxicological effects of organotin compounds (OTC), mainly tributyltin (TBT) and triphenyltin (TPhT) but also their di- and monosubstituted degradation products are well documented. Nowadays, the release of TBT from antifouling paints is recognized worldwide as being one of the main contamination problems for the marine environment, and the use of TBT-based antifouling paints is almost everywhere restricted by law. In order to evaluate the environmental distribution and fate of these compounds, and to control the effectiveness of these legal provisions, many analytical methods have been developed among which gas chromatography coupled to inductively coupled plasma mass spectrometry (GC-ICP-MS) is the most powerful.

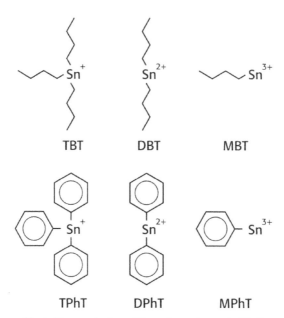

Fig. 2. The ionic form of butyltins and phenyltins of varying substitution.

TBT-based paints are used on vessels hulls to prevent growth of aquatic organisms that create roughness giving rise to increased drag, resulting in reduced vessel speed per unit energy consumption. The antifouling paint consists of a film-forming material with a biocidal ingredient and a pigment. It works by releasing small amounts of the biocide from the painted hull into the water, forming a thin envelope of highly concentrated TBT around the boat. The toxic concentration repels the settling stages of fouling organisms, like barnacles, seaweeds, or tubeworms on the boat's water-immersed surfaces [6].

However, the constant release of OTC from anti-fouling paints has led to toxic effects for nontarget aquatic species in the aquatic environment, where they cause deleterious effects, such as shell anomalies in oysters and imposex in gastropods, even at concentrations as low as nanograms per liter. Despite the restrictions organotin compounds (mainly TBT) are still used in paint formulation for large vessels, and about 69 % of all large ships are reported to use them [7]. Once released into the aquatic environment, organotin compounds may undergo a variety of degradation reactions until they are finally adsorbed onto suspended solids and sediments. Sediments are considered to be the ultimate sinks for organotin compounds [8].

In water, TBT decomposes into less toxic DBT and MBT species. The problem is that this favourable decomposition takes place far more slowly in sediments, creating an ecotoxicological risk long after its release into a given area, making sediments a secondary source of pollution. TBT and its degradation products DBT and MBT have been detected in different environmental compartments, both marine (waters, sediments, and biota) and terrestrial (waters and soils). The occurrences of the less toxic MBT and DBT compounds in the environment have so far been related to the degradation of TBT caused by microbial activity and/or photochemical reactions, but recently evidence for direct input of MBT and DBT was found. In fact the major application of organotin compounds (about 70 %) is the use of mono and dialkyltin derivatives as heat and light stabilizer additives in PVC processing [6].

3. Speciaition analysis of OTCs

Several techniques, based on species specific analytical methods, have been developed for the determination of butyltins in environmental matrices. Hyphenated systems, based on on-line coupling of gas chromatography (GC), liquid chromatography (LC), or supercritical fluid chromatography (SFC), to mass spectrometry (MS), inductively coupled plasma mass spectrometry (ICP-MS), atomic absorption spectrometry (AAS), and microwave induced plasma atomic emission spectrometry (MIP-AES) are in current use [7]. Among the different techniques, the coupling of GC to ICP-MS appears to be one of the most popular techniques, due to high sensitivity and multi-elemental and multi-isotopic capabilities [9]. Some of the most frequently used sample pretreatment methods for the determination of OTC, in various samples, are summarized in Table 1.

The coupling of chromatography with ICP-MS and application of calibration methods that are based on ID permits detection of several compounds. ID is a technique based on isotope ratio measurement whereby the natural isotopic abundance ratio of an analyte is altered by spiking with a standard that has a different isotopic abundance ratio. The prerequisite for the technique is that the analyte of interest should have more than one stable isotope [1]. In the case of tin (10 isotopes) the isotope of highest abundance is ^{120}Sn, usually referred as the reference isotope, and the spike isotope is generally one of the less abundant natural isotopes. For the purposes of speciation analysis, where OTCs are to be determined, there is a requirement for the isotopically enriched element-species to be synthesized. If two interference-free isotopes of a given element can be found, isotope dilution ICP-MS (ID-ICP-MS) can be performed, which generally provides superior accuracy and precision over other calibration strategies, including external calibration and the method of standard addition, because a ratio rather than an absolute intensity measurement is used in the quantification of the analyte concentration. Once equilibration is achieved between the analyte in the sample and the added spike, ID-ICP-MS is theoretically capable of compensating for any subsequent loss of analyte during sample manipulation, suppression of ion sensitivities by concomitant elements present in the sample matrix, and instrument drift. ID-ICP-MS may therefore be considered as a primary method of analysis and can play a crucial role for quality assurance in trace element chemical speciation of environmental and biological samples. ID-ICP-MS coupled with gas chromatography (GC) ID-GC-ICP-MS [9] and to less extent liquid chromatography (LC) ID-LC-ICP-MS [1] are the techniques that are used for speciation of organotin compounds. The derivatisation step is not needed in the case of LC (omitting one source of error), but the coupling with GC is superior when it comes to analysing samples of very low concentrations (pg/g) of OTC.

The presently available techniques for the determination of OTCs involve several analytical steps such as extraction, pre-concentration, cleanup, derivatisation (when gas chromatography is used), separation, and finally detection by element- or molecule specific techniques such as ICP-MS. The multitude of analytical steps is causing errors at various levels, making speciation analysis a difficult task.

4. Derivatization of OTCs for gas chromatographic separation

In order to enable separation by GC, the ionic organotin compounds need to be converted into volatile species such as the hydrides, or their fully alkylated form (derivatisation). For the reduction to hydrides, sodium tetrahydroborate ($NaBH_4$) is commonly used, whereas derivatisation through alkylation can be carried out with Grignard reagents, sodium tetraethylborate ($NaBEt_4$), or sodium tetra (n-propyl) borate ($NaBPr_4$). Derivatisation with $NaBEt_4$ makes the sample preparation faster and easier because it combines an *in situ* derivatisation with extraction of the ethylated organotin compounds into an organic phase [1].

Sample type	Sample pretreatment	Compound studied (amount found (ng))	Separation and detection technique	Ref
Marine sediment from contaminated harbour area in Mar piccolo (Italy)	Add 3 mL HCl and 6 mL methanol to 1 g of sample. Shake and sonicate for 15 min in ultrasonic bath. Add 3 mL of acetate buffer (pH 5.3) and centrifuge the leached sediment at 4000 rpm for 15 min. Pipette 1 mL supernatant into a 15 mL glass vial and add 6 mL of HOAc/NaOAc buffer (pH 5.3). Close the vial with a septum and add 1 mL of $NaBEt_4$ solution with a syringe. Sonicate the reaction mixture and pierce the SPME needle into the septum and expose the fiber into the headspace.	MBT = 8 DBT = 10 TBT = 1	Headspace micro-extraction GC/MS	28
Sediment samples Samples 1 and 2 from the harbour of Ostend, sample 3 from a dry dock in the harbour of Antwerp, and sample 4 from a leisure craft maintenance place located in the province of Limburg, all in Belgium	Add 2 ml of HCl (32 %) and 8 ml H_2O to 1 g sample in a centrifuga-tion vessel. Add 25 ml of hexane-ethyl acetate mixture (1:1) contain-ning 0.05 % tropolone. Sonicate the mixture for 1 h, followed by centrifugation at 3000 rpm for 5 min. Transfer the organic phase in-to an extraction vessel and evaporate to dryness using rotary evaporation. Add 0.5 zml of hexane containing Pe_3SnEt as an internal standard and derivatize by adding 1 ml of $NaBEt_4$ solution together with 50 ml of acetate buffer solution. Shake the mixture manually for 5 min	**Sample 1** MBT= 0.14 ± 0.02 DBT= 0.44 ± 0.03 TBT= 0.14 ± 0.02 **Sample 2** MBT= 0.36 ± 0.02 DBT= 1.11 ± 0.12 TBT= 2.33 ± 0.14 **Sample 3** MBT= 8.13 ± 0.33 DBT= 10.0 ± 0.6 TBT= 26.4 ± 1.8 **Sample 4**	GC interfaced with AAS and AES	14

Sample type	Sample pretreatment	Compound studied (amount found (ng))	Separation and detection technique	Ref
	and seperate the hexane phase. Introduce the extract into a clean-up column (a pasteur pipette filled with alumina to form a plug of 5 cm). Add an additional volume of 1 ml diethyl ether and evaporate the added diethyl ether using a gentle stream of nitrogen.	MBT= 1.55 ± 0.09 DBT= 1.67 ± 0.20 TBT= 6.60 ± 0.18		
Sediment samples Two samples (1&2) from the harbour of Ostend (Belgium).	Add 4 ml H_2O, 1 ml acetic acid (96 %), 1 ml DDTC in pentane, and 25 ml hexane into a 100 ml Erlenmeyer flask containing 1 g sample. Sonicate the mixture for 30 min and decant the organic phase into a 100 ml beaker. Repeat the extraction with 25 ml of hexane and stir magnetically for 30 min. Centrifuge the mixture for 5 min at 3000 rpm. Dry over Na_2SO_4 and evaporate to dryness on a rotary evaporator. Add 250 µl of n-octane, containing Pr_3SnPe and pentylate with 1 ml of 1 M n-PeMgBr. Destroy excess Grignard reagent by adding 10 ml of 0.5 M H_2SO_4. Introduce the octane layer into a clean-up column (a pasteur pipette filled with alumina to form a plug of 5 cm). Add an additional volume of 1 ml diethyl ether and evaporate the added diethyl ether using a gentle stream of nitrogen.	**Sample 1** DBT= 0.43 ± 0.02 TBT= 0.31 ± 0.03 **Sample 2** DBT= 1.39 ± 0.06 TBT= 2.67 ± 0.08	GC interfaced with AAS and AES	14
Water 1. Sea water from Sahrm el Sheikh harbour in South Sinai (Egypt) 2. Harbour water from Wädenswil, Lake Zurich (Switzerland)	Add 0.5 mL of acetic acid/acetate buffer solution (pH 5) and 1.45 g of sodium chloride to 50 mL water sample. After shaking spike with 100 µl of deuterated standard solution mixture (12.5 ng/L of each species in MeOH). Shake and add 150 µl of 1.5 % (w/v) NaBEt₄ aqueous solution. Add 1 ml of hexane and shake in the dark for	**1. Sahrm el Sheikh** MBT = 3.4 ± 7.6 DBT = 2.1 ± 24 TBT = 2.6 ± 17 MPhT= 1.5 ± 6.7 DPhT = 0.5 ± 76 TPhT = 4.8 ± 2.1 **2. Wädenswil** MBT = nd	LLE, Large volume injection GC/MS	29

Sample type	Sample pretreatment	Compound studied (amount found (ng))	Separation and detection technique	Ref
	12 h at 25 °C. Transfer 180 µl of the hexane extract to a 1 mL auto-sampler vial and spike with 10 µl of 0.2 ng/µl TeBT in hexane. Inject 50 µl in to the GC.	DBT = 3 ± 17 TBT = nd MPhT = 37 ± 3 DPhT = 23 ± 6 TPhT = 353 ± 3		
Sediment Certified reference material PACS-2 (0.98 ± 0.13 TBT ng/L)	Put 0.5 g of PACS-2 in a Prolabo microwave digester and spike with 0.04 mL of ^{117}Sn-enriched TBT solution. Add 5 mL of acetic acid and heat at 60 % power for 3 min. Centrifuge at 2000 rpm for 5 min. Transfer 1 mL volume of the supernatant to a reaction vial and add 1 mL of deionized water. Adjust to pH 5-6 with 1.2 mL ammonium hydroxide. Buffer the content with 0.8 mL of ammonium citrate (2 mol L^{-1}) and dilute to 10 mL with methanol.	TBT = 1.018 ± 0.0315	Microwave extraction, ID-HPLC-ICP-MS	30
Sediment/Sludge 1. Sediment pore water from (1-20 cm) from Stansstaad harbour, Lake Lucerne (Switzerland) 2. Sewage sludge from four wastewater treatment plants in the Zurich canton (Switzerland)	Weigh 2.5 g freeze dried sediment or sludge in a beaker. Spike homogeneously with 500 µl of deuterated standard solution mixture (12.5 ng/L of each species in MeOH). Mix with 9 g of quartz sand and transfer the mixture to 11 mL extraction cells. After two hours fill the extraction cells with a mixture of 1 M sodium acetate and 1 M acetic acid in MeOH, using ASE. Extract with three to five static cycles of 5 min. Renew 4 mL of solvent between each static extraction. Rinse the cells with 4 mL of solvent and purge with nitrogen. Transfer the combined extracts to 250 mL volumetric flasks containing 7.3 g of NaCl. Add water and adjust the pH to 5 with 1 M NaOH. Add 1 mL of aqueous solution of 5 % (w/v) NaBEt$_4$ and fill the bottles to 250 mL with water. Add 2	1) Sediment pore water MBT = 11.0± 1.9 DBT = 4.5 ± 4.7 TBT = 9.6 ± 4.4 MPhT = 12.3 ± 4 DPhT = 2.5 ± 14 TPhT = 4.1± 3 2) Sewage sludge MBT = 300 ± 4 DBT = 253 ± 5 TBT = 45 ± 5 MPhT = 7 ± 21 DPhT = nd TPhT = 7 ± 38	ASE, Large volume injection GC/MS	29

Sample type	Sample pretreatment	Compound studied (amount found (ng))	Separation and detection technique	Ref
	mL of hexane and shake for 12 h. Transfer 500 µl of the hexane extract to 2 mL GC vials and spike with 10 µl of TeBT (5 ng/µl). For sewage sludge transfer the hexane extract to 10 mL centrifuge tubes containing 0.9 g of deactivated silica gel and 2 mL water. Shake vigorously and centrifuge.			

Table 1. Some of the most commonly used methods for the determination of OTCs in various samples along with the analysis results obtained.

5. Species Specific Isotope Dilution (SSID)

In the last few years different procedures for the speciation of organotin compounds have been proposed, among which species specific isotope dilution (SSID) is one [10-11]. The use of isotope enriched spike solutions has not only the potential for accurate, precise and simultaneous determination of OTCs but also the evaluation of different extraction and derivatization protocols for the analysis of sediments and biological sample in which species interconversion/decomposition can take place during sample preparation. Organotin compounds synthesized from isotopically enriched tin metals are used for preparing spikes that can be used for the determination of concentrations using species specific isotope dilution (SSID). Different types of calibration strategies based on SSID have been used for the determination of organotin compounds:

i.   Single isotope - species specific isotope dilution (SI-SSID): where a mixture containing organotin species, all with the same isotope, is used as a spike. For example: a mixture containing ^{116}Sn-enriched MBT, ^{116}Sn-enriched DBT, and ^{116}Sn-enriched TBT is used for the determination of butyltins.

ii. Multiple isotope - species specific isotope dilution (MI-SSID): where a mixture containing the organotin species, each with different isotope, is used as a spike. For example: a mixture containing ^{119}Sn-enriched MBT, ^{118}Sn-enriched DBT, and ^{117}Sn-enriched TBT is used for the determination of butyltins.

Single isotope species specific isotope dilution (SI-SSID) has been used during the last fifteen years for speciation analysis of bromine [12], chromium [13], iodine [14], selenium [15,16], mercury [17] and tin [18-20]. Recently multi-isotope species specific isotope dilution (MI-SSID) has been used to monitor and correct for the degradation/transformation that takes place during sample preparation [21-23]. The use of multi isotope spike can also be extended to compare different extraction procedures and study the inherent procedural parameters that govern the process of degradation/transformation. If these problems are fully understood and addressed, appropriate strategies can be set to facilitate accurate speciation analysis of the compounds of interest in various types of samples.

One of the drawbacks with SI-SSID is the difficulty to assess the degradation/redistribution of the OTs that can take place during sample workup. Another problem is the inability to match the concentrations of the organotin compounds in the spike with those present in the sample, thus affecting the accuracy and precision of the results [18, 24-26]. The latter problem can however be circumvented by using MI-SSID [5, 23].

Owing to the problems with degradation of phenyltin species during the sample pretreatment steps, poor precisions within and large spreads between laboratory results were obtained when certifying organotin compounds in the BCR CRM-477. As a consequence, only indicative values were given for phenyltin species [27]. Figure 3 shows deconvoluted chromatograms obtained for the multi-isotope standard where no peaks, corresponding to redistribution products of the enriched phenyltin standard, are visible. This indicates that the integrity of the phenyltin species is preserved during the ethylation and the detection procedure.

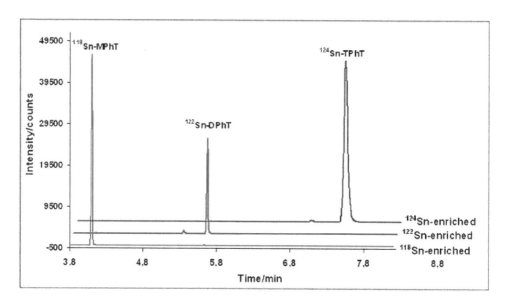

Fig. 3. Deconvoluted GC-ICPMS sub-chromatograms obtained for a mixture of [118]Sn-enriched MPhT (106.5 ng g[-1]), [122]Sn-enriched DPhT (57.4 ng g[-1]) and [124]Sn-enriched TPhT (239.0 ng g[-1]). The chromatograms are shifted from each other by 0.1 minute and 3000 counts for clarity. [*From Van D.N., Muppala S.R.K., Frech W., Tesfalidet S., Preparation, preservation and application of pure isotope-enriched phenyltin species Anal. Bioanal. Chem., 2006, 386, 1505*]

The presently available techniques for the determination of OTCs involve several analytical steps such as extraction, pre-concentration, cleanup, derivatisation (when gas chromatography is used), separation, and finally detection by element- or molecule specific techniques such as ICP-MS. The multitude of analytical steps is causing errors at various levels, making speciation analysis a difficult task.

In the determination of PhTs, species transformation/interconversion is one of the most serious obstacles which affects the reliability of the analytical results. Using MI-SSID, one can assess the degradation/transformation processes, enabling calculation of degradation-corrected concentrations for the species of interest. Possible degradation/transformation pathways of PhTs are depicted in Fig. 4

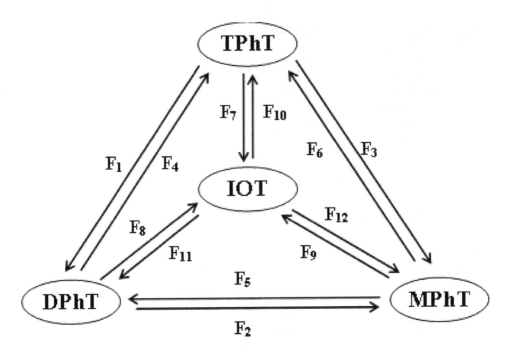

Fig. 4. Possible interconversion pathways of inorganic tin (IOT) and PhTs. F_i is interconversion factor corresponding to the interconversion reaction i. TPhT triphenyltin, DPhT diphenyltin, MPhT monophenyltin. [*From Van D.N., Bui T.T.X., Tesfalidet S. "The transformation of phenyltin species during sample preparation of biological tissues using multi-isotope spike SSID-GC-ICPMS, Anal. Bioanal. Chem 2008, 392, 737*]

6. Instrumentation and operating conditions for the GC and ICP-MS

A Varian 3300 gas chromatograph (Palo Alto, CA, USA) fitted with an on-column injector liner and a methyl silicone capillary column (30 m by 0.53 mm i.d., 1.5 μ film thickness; SPB-1, Supelco, Bellafonte, PA) was used for separation of Sn species and an Agilent 7500 ICP-MS (Foster City, CA) was used for detection. The GC was coupled to the ICP-MS *via* a custom made interface. The operating parameters of the ICP-MS were selected by optimizing the sensitivity for [129]Xe, by adding Xe gas at 0.5 ml/min to the Ar carrier gas flow. Oxygen was added to the plasma to prevent carbon deposits on the sampler/skimmer Pt cones. The operating parameters of the GC and ICP-MS are given in Table 2.

GC parameters	
Injection volume	1 μL
Carrier helium gas flow	22 mL min⁻¹
Injector temperature	180 °C
Oven temperature	130 °C ⊕ 40 °C min⁻¹ ⊕ 210 °C, hold 0.5 minutes
	210 °C ⊕ 7 °C min⁻¹ ⊕ 225 °C, hold 0.5 minutes
	225 °C ⊕ 40 °C min⁻¹ ⊕ 280 °C, hold 2.0 minutes
Transfer line temperature	200 °C
ICP-MS parameters	
ICP RF power	1200 W
Plasma argon gas flow	15 L min⁻¹
Nebulizer argon gas flow	1.0 L min⁻¹
Auxiliary argon gas flow	0.9 L min⁻¹
Auxiliary oxygen gas flow	3 mL min⁻¹
Sampler/Skimmer cones	Platinum
Dwell time	100 ms for $^{116}Sn^+$, $^{117}Sn^+$, $^{118}Sn^+$, $^{119}Sn^+$, $^{120}Sn^+$ and $^{124}Sn^+$

Table 2. Operating conditions for the GC and ICP-MS.

7. Determination of OTCs in biological samples

Various extraction procedures have been used for the analysis of organotin compounds in biological samples. Pellegrino et al. [31] compared twelve selected extraction methods to evaluate the extraction efficiencies obtained for a certified reference material (CRM). The organic solvent, the nature and concentration of the acid used for leaching, and the presence/absence of tropolone used as complexing agent, were all found to influence the grade of transformation that takes place during extraction [32]. Although the performance of the extraction procedures can be validated by recovery tests using certified reference materials (CRMs) or fortified samples, the validity of these procedures on real samples cannot be guaranteed [18,33]. This is because the adsorption/binding forces of the species of interest to the solid are strongly dependent on the matrix [10]. A mild extraction technique could, for example, give rise to incomplete extraction of organotin species from the solid sample while a harsher extraction technique facilitates the degradation/rearrangement of the species of interest [32,34,35]. In both cases, the results of speciation analysis will fail to reflect the real speciation in the solid sample. As a result, it has yet not been able to certify the concentration of phenyltins in the reference material BCR CRM-477 and only indicative values are available [36].

8. Analysis of PhTs in mussel tissue

When analyzing PhTs in certified reference material, mussel tissue, BCR CRM-477, 8.7±0.9% of TPhT was converted to DPhT and 9.0±0.2% of DPhT to MPhT. Other dephenylation/phenylation products were not observed. Peaks indicating the degradation of ^{124}Sn-enriched TPhT and ^{122}Sn-enriched DPhT are shown in Fig. 5 by signals for ^{124}Sn-enriched DPhT and ^{122}Sn-enriched MPhT [5].

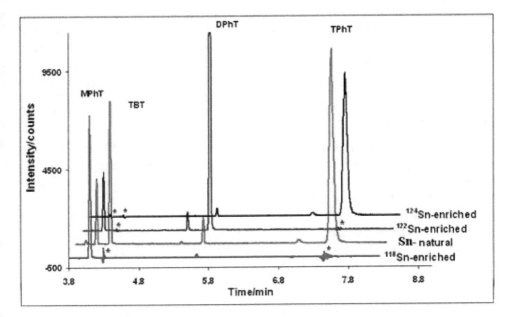

Fig. 5. Deconvoluted GC-ICPMS sub-chromatograms obtained for 0.2g BCR CRM-477 spiked with 0.14g [118]Sn-enriched MPhT (90.2 ng g[-1]), [122]Sn-enriched DPhT (367 ng g[-1]) and [124]Sn-enriched TPhT (315 ng g[-1]). (*) is the fluctuation of the baseline due to deconvolution. The chromatograms are shifted from each other by 0.1 minute and 700 counts for clarity.
[*From Van D.N., Muppala S.R.K., Frech W., Tesfalidet S., Preparation, preservation and application of pure isotope-enriched phenyltin species Anal. Bioanal. Chem., 2006, 386, 1505*]

8.1 Equations for calculating the inter-conversion factors and concentrations of PhT species

Various equations have been proposed for the calculation of concentrations in isotope dilution [17, 37-38]. The system of equations used for calculating the inter-conversion factors and concentrations of PhT species presented here are similar to those presented by Rodriguez-Gonzalez et.al. [37] and Nguyen D.N. et.al [5].

A deconvoluted chromatogram of the type shown in figure 5 visualizes how each added isotope-enriched species is distributed between the PhT species. Deconvolution is performed for each species by solving the system of four linear equations (Eq. 4) in an Excel spreadsheet:

$$\sum S_{118} = R_{118/118} \cdot S_{118} + R_{118/120} \cdot S_{120} + R_{118/122} \cdot S_{122} + R_{118/124} \cdot S_{124}$$
$$\sum S_{120} = R_{120/118} \cdot S_{118} + R_{120/120} \cdot S_{120} + R_{120/122} \cdot S_{122} + R_{120/124} \cdot S_{124}$$
$$\sum S_{122} = R_{122/118} \cdot S_{118} + R_{122/120} \cdot S_{120} + R_{122/122} \cdot S_{122} + R_{122/124} \cdot S_{124}$$
$$\sum S_{124} = R_{124/118} \cdot S_{118} + R_{124/120} \cdot S_{120} + R_{124/122} \cdot S_{122} + R_{124/124} \cdot S_{124}$$

(4)

To determine the degree of inter-conversion between PhTs, the isotope ratios 120/118, 120/122 and 120/124 for each PhTs were calculated. The following inter-conversion factors were considered: F_1; TPhT degradation to DPhT, F_2; DPhT to MPhT and F_3; TPhT to MPhT F_4; DPhT phenylation to TPhT, F_5; MPhT to DPhT and F_6; MPhT to TPhT. The calculations of two inter-conversion factors related to the formation of TPhT (F_4 and F_6) are exemplified in equations 5&6. Once all the inter-conversion factors are calculated, mass balance equations for each species can be established and used for computing concentrations of each species in the sample. For example, the third row in the matrix (equation 8) is obtained by rearranging Equation 4, the mass balance of TPhT.

The symbols used in the equations represent the following: N_s^{MPhT}, N_s^{DPhT} and N_s^{TPhT} are the number of moles of each species present in the sample; N_{sp}^X is the number of moles of the PhTs X (X = MPhT, DPhT or TPhT) spiked into the sample; At_s^i is the abundance of tin isotope i (i = 118, 120, 122 or 124) in the sample; $At_{i,sp}^{MPhT}$, $At_{i,sp}^{DPhT}$ and $At_{i,sp}^{TPhT}$ are the abundance of tin isotope i in the spiked ^{118}Sn-enriched MPhT, ^{122}Sn-enriched DPhT and ^{124}Sn-enriched TPhT, respectively. $N_{124,m}^X$, $N_{122,m}^X$, $N_{120,m}^X$ and $N_{118,m}^X$ are the number of moles of the corresponding isotopic PhTs X in the blend; $R_{j/k,m}^X$ is the tin isotope ratio of the reference isotope j (j = 120) to the spiked isotope k (k = 118, 122, 124) measured by GC-ICPMS for the phenyltin X in the blend after mass bias correction.

$$N_{sp}^{TPhT}\left(\frac{At_{120,sp}^{TPhT} - R_{120/124,m}^{TPhT}At_{124,sp}^{TPhT}}{R_{120/124,m}^{TPhT}At_{124,s} - At_{120,s}} - \frac{At_{120,sp}^{TPhT} - R_{120/122,m}^{TPhT}At_{122,sp}^{TPhT}}{R_{120/122,m}^{TPhT}At_{122,s} - At_{120,s}}\right)$$
$$+F_4 N_{sp}^{DPhT}\left(\frac{At_{120,sp}^{DPhT} - R_{120/124,m}^{TPhT}At_{124,sp}^{DPhT}}{R_{120/124,m}^{TPhT}At_{124,s} - At_{120,s}} - \frac{At_{120,sp}^{DPhT} - R_{120/122,m}^{TPhT}At_{122,sp}^{DPhT}}{R_{120/122,m}^{TPhT}At_{122,s} - At_{120,s}}\right) \quad (5)$$
$$+F_6 N_{sp}^{MPhT}\left(\frac{At_{120,sp}^{MPhT} - R_{120/124,m}^{TPhT}At_{124,sp}^{MPhT}}{R_{120/124,m}^{TPhT}At_{124,s} - At_{120,s}} - \frac{At_{120,sp}^{MPhT} - R_{120/122,m}^{TPhT}At_{122,sp}^{MPhT}}{R_{120/122,m}^{TPhT}At_{122,s} - At_{120,s}}\right) = 0$$

$$N_{sp}^{TPhT}\left(\frac{At_{120,sp}^{TPhT} - R_{120/124,m}^{TPhT}At_{124,sp}^{TPhT}}{R_{120/124,m}^{TPhT}At_{124,s} - At_{120,s}} - \frac{At_{120,sp}^{TPhT} - R_{120/118,m}^{TPhT}At_{118,sp}^{TPhT}}{R_{120/118,m}^{TPhT}At_{118,s} - At_{120,s}}\right)$$
$$+F_4 N_{sp}^{DPhT}\left(\frac{At_{120,sp}^{DPhT} - R_{120/124,m}^{TPhT}At_{124,sp}^{DPhT}}{R_{120/124,m}^{TPhT}At_{124,s} - At_{120,s}} - \frac{At_{122,sp}^{DPhT} - R_{120/118,m}^{TPhT}At_{118,sp}^{DPhT}}{R_{120/118,m}^{TPhT}At_{118,s} - At_{120,s}}\right) \quad (6)$$
$$+F_6 N_{sp}^{MPhT}\left(\frac{At_{120,sp}^{MPhT} - R_{120/124,m}^{TPhT}At_{124,sp}^{MPhT}}{R_{120/124,m}^{TPhT}At_{124,s} - At_{120,s}} - \frac{At_{120,sp}^{MPhT} - R_{120/118,m}^{TPhT}At_{118,sp}^{MPhT}}{R_{120/118,m}^{TPhT}At_{118,s} - At_{120,s}}\right) = 0$$

$$F_6 N_s^{MPhT} + F_4 N_s^{DPhT} + N_s^{TPhT} = N_{sp}^{TPhT}\frac{At_{120,sp}^{TPhT} - R_{120/124,m}^{TPhT}At_{124,sp}^{TPhT}}{R_{120/124,m}^{TPhT}At_{124,s} - At_{120,s}}$$
$$+F_4 N_{sp}^{DPhT}\frac{At_{120,sp}^{DPhT} - R_{120/124,m}^{TPhT}At_{124,sp}^{DPhT}}{R_{120/124,m}^{TPhT}At_{124,s} - At_{120,s}} + F_6 N_{sp}^{MPhT}\frac{At_{120,sp}^{MPhT} - R_{120/124,m}^{TPhT}At_{124,sp}^{MPhT}}{R_{120/124,m}^{TPhT}At_{124,s} - At_{120,s}} \quad (7)$$

$$\begin{pmatrix} 1 & F_2 & F_3 \\ F_5 & 1 & F_1 \\ F_6 & F_4 & 1 \end{pmatrix} \begin{pmatrix} N_s^{MPhT} \\ N_s^{DPhT} \\ N_s^{TPhT} \end{pmatrix} = \begin{pmatrix} X_{MPhT} \\ Y_{DPhT} \\ Z_{TPhT} \end{pmatrix} \qquad (8)$$

Solving Matrix 8 by setting the F values (both on the left side and even included in X_{MPhT}, Y_{DPhT}, and Z_{TPhT}) equal to zero will correspond to degradation-uncorrected concentrations of the phenyltins.

9. References

[1] Hill S.J., Pitts L.J., Fisher A.S. *Trends Anal. Chem.*, 2000, 19, 120.

[2] Berglund M., 'Introduction to Isotope Dilution Mass Spectrometry (IDMS)', *Handbook of Stable Isotope Analytical Techniques*, ed. P. De Groot, Elsevier, Amsterdam, 2004, vol. 1, 820–834.

[3] Jackson C.H.N. *J. Anim. Ecol.*, 1933, 2, 204.

[4] Vogl J. *J. Anal. At. Spectrom.*, 2007, 22, 475.

[5] Nguyen V.D., Bui T.T., Tesfalidet S., *Anal. Bioanal. Chem.*, 2008, 392, 737.

[6] Hoch M., *Applied Geochemistry*, 2001, 16, 719.

[7] Tao H., Rajendran R. B., Quetel C. R., Nakazato T., Tominaga M., Miyazaki A., *Anal. Chem.*, 1999, 71, 4208.

[8] Schubert P., Rosenberg E., Grasserbauer M. *Intern. J. Environ. Anal. Chem.*, 2000, 78, 185.

[9] Encinar J. R., Alonso J. I. G., Sanz-Medel A., *J. Anal. Atom. Spectrom.*, 2000, 15, 1233.

[10] Berg M., Arnold C. G., Muller S. R., Muhlemann J., Schwarzenbach R. P., *Environ. Sci. Technol.*, 2001, 35, 3151.

[11] Arnold C. G., Berg M., Muller S. R., Dommann U., Schwarzenbach R. P., *Anal. Chem.*, 1998, 70, 3094.

[12] Diemer J, Heumann KG., *Fresenius J. Anal. Chem.*, 1997, 357, 74–79.

[13] Tirez K, Brusten W, Cluyts A, Patyn J, De Brucker N., *J. Anal. Atom. Spectrom.*, 2003, 18, 922–932.

[14] Reifenhäuser C, Heumann K.G., *Fresenius J. Anal. Chem.*, 1990, 336, 559.

[15] Minami H., Cai W.T., Kusumoto T., Nishikawa K., Zhang Q., Inoue S., Atsuya I., *Anal. Sci.*, 2003, 19, 1359.

[16] Hinojosa R. L., Marchante G.J.M., Alonso J.I.G., Sanz-Medel A., *J. Anal. At. Spectrom.*, 2004, 19, 1230.

[17] Hintelmann H, Evans R.D., *Fresenius J. Anal. Chem.*, 1997, 358:375.

[18] Kumar S.J., Tesfalidet S., Snell J.P., Van D.N., Frech W., *J. Anal. At. Spectrom.*, 2004, 19, 368.

[19] Kumar S.J., Tesfalidet S., Snell J., Frech W., J., *Anal. At. Spectrom.*, 2003, 18, 714.

[20] Encinar J.R., Villar M.I.M, Santamaria V.G., Alonso J.I.G., Sanz-Medel A., *Anal. Chem.*, 2001, 73, 3174.

[21] Qvarnström J., Frech W., *J. Anal. At. Spectrom.*, 2002, 17, 1486.

[22] Encinar J.R, Rodríguez-Gonzalez P, Alonso J.I.G., Sanz-Medel A., *Anal. Chem.*, 2002, 74, 270.

[23] Van D.N., Muppala S.R.K., Frech W., Tesfalidet S., *Anal. Bioanal. Chem.*, 2006, 386, 1505.

[24] Dauchy X., Cottier R., Batel A., Jeannot R., Borsier M., *J. Chromatogr. Sci.*, 1993, 31, 416.

[25] Kumar S.J., Tesfalidet S., Snell J., Frech W. *J. Anal. At. Spectrom.*, 2003, 18, 714.

[26] Tesfalidet S. Screening of organotin compounds in the Swedish environment, Final report; SNV contract: 219 0102, 2004. http://www.ivl.se/miljo/projekt/ dvss/pdf/organotenn.pdf

[27] Morabito R, Soldati P., de la Calle M.B., Quevauviller Ph., *Appl. Organomet. Chem.*, 1998, 12, 621.

[28] Cardellicchio N., Giandomenico S., Decataldo A., Di Leo A., *Fresenius J. Anal. Chem.*, 2001, *369*, 510.

[29] Ceulemans M., Adams F. C., *Anal. Chim. Acta.*, 1995, *317*, 161.

[30] Yang L., Mester Z., Sturgeon R. E., *Anal. Chem.*, 2002, *74*, 2968.

[31] Pellegrino C., Massanisso P. , Morabito R., *Trends Anal. Chem.*, 2000, 19, 97.

[32] Abalos M. , Bayona J.M. , Quevauviller Ph., *App., Organomet. Chem*, 1998, 12, 541.

[33] Kumar S.J., Tesfalidet S., Snell J., Frech W., *J. Anal. At. Spectrom.* 2003, 18, 714.

[34] Ariese F., Cofino W., Gomez-Ariza J.L., Kramer G.N., Quevauviller Ph., *J. Environ. Monit.* 1999, 2, 191.

[35] Alonso I.J.G., Encinar J., González R.P., Sanz-Medel A., *Anal. Bioanal. Chem.* 2002, 373, 432.

[36] Morabito R., Soldati P., de la Calle Gutinas M.B., Quevauviller Ph., *Appl. Organomet. Chem.* 1998, 12, 621.

[37] Gonzalez P.R., Alonso J.G., Sanz-Medel A., *J. Anal. At. Spectrom.* 2005, 20, 1076.

[38] Meija J., Yang L., Caruso J.A., Mester Z., *J. Anal. At. Spectrom.* 2006, 21, 1294.

Part 3

New Techniques in Gas Chromatography

9

Inverse Gas Chromatography in Characterization of Composites Interaction

Kasylda Milczewska and Adam Voelkel
Poznan University of Technology,
Institute of Chemical Technology and Engineering, Poznan,
Poland

1. Introduction

Inverse gas chromatography is a useful and quite versatile technique for materials' characterization, because it can provide information on thermodynamic properties over a wide temperature range. The term "inverse" indicates that the stationary phase of the chromatographic column is of interest, in contrast to conventional gas chromatography. The chromatographic column contains the material under study. The method is simple, fast and efficient. It has been used for the characterization of hyperbranched polymers [Dritsas et al., 2008], block copolymers [Zou et al., 2006], polymer blends [Al-Ghamdi & Al-Saigh, 2000], nanocomposites [Boukerma et al., 2006], fillers [Milczewska & Voelkel, 2002], cement pastes [Oliva et al., 2002], fibers [van Asten et al., 2000] and crude oils [Mutelet et al., 2002].

Mixtures of different types of materials i.e. polymers, blends, modified fillers or compositions are utilized extensively to produce commercially useful materials having combinations of properties not revealed by a single component. Many of the properties and processing characteristics of those mixtures depend on whether they are miscible or not. Theory operates with parameters relating to the pure components [Voelkel et al., 2009 (a)]. The knowledge of the interaction parameters between polymers and solvents is very important in the study of their miscibility and thermodynamic properties of solutions [Huang, 2009].

The interactions between one probe and the polymer are usually characterized by the values of Flory–Huggins interaction parameter [Dritsas et al., 2009]. Only a few techniques can provide quantitative information about the change of free energy when mixing two components. The data from P-V-T experiments might be successfully used in the prediction of the miscibility of polyolefine blends [Han et al., 1999]. Interaction parameter for the components of polymer blends was also determined with the use of small angle x-ray scattering (SAXS) [de Gennes, 1979; Meoer & Strobl, 1987; Ying et al. 1993], thermal induced phase separation (TIPS) [Sun et al., 1999] and small angle neutron scattering (SANS) [Fernandez et al., 1995; Hindawi et al., 1990; Horst & Wolf, 1992; Mani et al., 1992; Schwann et al., 1996]. In last two decades, the Flory-Huggins interaction parameter was also determined using the melting point depression method for crystal-containing polymers by differential scanning calorimetry (DSC) [Lee et al., 1997]. It is worth to note the increasing role of inverse gas chromatography (IGC) [Voelkel et al., 2009 (a)], because of its simplicity, rapidity, and the general availability of GC equipment.

2. Theory of interaction

Inverse Gas Chromatography (IGC) is a gas-phase technique for characterizing surface and bulk properties of solid materials. The principles of IGC are very simple, being the reverse of a conventional gas chromatographic (GC) experiment.

While it is a dynamic method, it was shown many years ago that measurements recorded under the correct conditions could give accurate equilibrium thermodynamic information [Shillcock & Price, 2003]. The retention of a solvent or 'probe' molecule on the material is recorded and the measurement made effectively at infinite dilution of the probe. A range of thermodynamic parameters can then be calculated. One advantage of the method is that it is readily applied to mixtures of two or more polymers.

A cylindrical column is uniformly packed with the solid material of interest, typically a powder, fiber or film. A pulse or constant concentration of gas is then injected down the column at a fixed carrier gas flow rate, and the time taken for the pulse or concentration front to elute down the column is measured by a detector. A series of IGC measurements with different gas phase probe molecules then allows access to a wide range of physico-chemical properties of the solid sample [SMS-*i*GC brochure 2002].

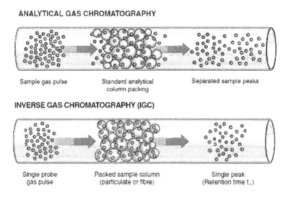

Fig. 1. Analytical vs. Inverse gas chromatography

When a liquid probe is injected into the column, the probe vaporizes and flows with the carrier gas, and a characteristic specific retention volume (V_g) can be measured:

$$V_g = \frac{3}{2} \cdot \frac{t_R^{'} \cdot j \cdot F \cdot 273.15}{m_w \cdot T} \quad (1)$$

where: $t_R^{'} = t_R - t_M$, t_M - gas hold-up time, calculated by Grobler–Balizs procedure [Grobler & Balizs, 1974], j – James-Martin's coefficient [James & Martin, 1952].

3. Flory-Huggins parameters

The properties of polymer blends are determined mainly by the miscibility of the components and structure. Usually thermodynamic miscibility and homogeneity can be attained when the free energy of mixing is negative. The classical thermodynamics of binary polymer–solvent systems was developed independently by P.J. Flory [Flory, 1942] and M.L. Huggins [Huggins, 1942]. It is based on the well-known lattice model qualitatively formulated by K.H. Meyer [Meyer, 1939], who pointed out the effect of the differences in molecular size of polymer and solvent molecules on the entropy of mixing. The quantitative calculation of the entropy of mixing led to the introduction of a dimensionless value, the so-called Flory-Huggins interaction parameter, for the thermodynamic description of polymer solutions [Gundert & Wolf, 1989]. Flory–Huggins interaction parameter (χ) is an important factor of miscibility of polymer blends and solutions.

Using Flory–Huggins theory, the Flory–Huggins interaction parameter between a polymer and probe, χ, can be related to the specific retention volume of probes, V_g, by the following equation [Milczewska & Voelkel, 2002 as cited in Barrales-Rienda, 1988; Voelkel et al., 2009 (b) as cited in Voelkel & Fall, 1995]:

$$\chi_{12}^{\infty} = \ln\left(\frac{273.15 \cdot R}{p_1^o \cdot V_g \cdot M_1}\right) - \frac{p_1^o}{R \cdot T} \cdot \left(B_{11} - V_1^o\right) + \ln\left(\frac{\rho_1}{\rho_2}\right) - \left(1 - \frac{V_1^o}{V_2^o}\right) \tag{2}$$

1 denotes the solute and 2 denotes examined material, M_1 is the molecular weight of the solute, p_1^o is the saturated vapor pressure of the solute, B_{11} is the second virial coefficient of the solute, V_i^o is the molar volume, ρ_i is the density, R is the gas constant.

This equation may be rearranged into form including weight fraction activity coefficient:

$$\ln \Omega_1^{\infty} = \ln\left(\frac{a_1}{w_1}\right) = \ln\left(\frac{273.15 \cdot R}{p_1^o \cdot V_g \cdot M_1}\right) - \frac{p_1^o}{R \cdot T} \cdot \left(B_{11} - V_1^o\right) \tag{3}$$

$$\chi_{12}^{\infty} = \ln \Omega_1^{\infty} + \ln\left(\frac{\rho_1}{\rho_2}\right) - \left(1 - \frac{V_1^o}{V_2^o}\right) \tag{4}$$

When the data of the density and the molecular mass of both the solute and the stationary phase (polymer) are inaccessible it is possible to determine the Flory-Huggins interaction parameter by simplifying Eq. (4):

$$\chi_{12}^{\infty} = \ln \Omega_1^{\infty} - 1 \tag{5}$$

i.e., under the assumption that $\ln\left(\frac{\rho_1}{\rho_2}\right) = 0$ which means that the densities of the solute and

the stationary phase are of similar order and $\frac{V_1^o}{V_2^o} \rightarrow 0$ (the molar volume of the stationary

phase is much higher than that of the test solute) [Voelkel & Fall, 1997].

Etxabarren et al. [Etxabarren et al., 2002] described molecular mass, temperature and concentration dependences of the polymer-solvent interaction parameter. The concentration dependence has been reasonably explained after the consideration of the different compressibilities (or free volumes) of the components. A parabolic dependence of χ with temperature is necessary in order to explain the lower or upper critical solution temperatures characteristic of most of the polymer solutions. In fact, there are experimental evidences of such type of dependence although because of limitations imposed by the degradation of the polymer and the freezing point of the solvent, a limited temperature range can be studied and only a part of this parabolic curve is usually evidenced.

Molecular mass dependence of the interaction parameter has been a recurrent subject in the polymer literature, and Petri et al. [Petri et al., 1995] have reported new experimental results which seem to indicate that there is a real molecular mass dependence of χ, especially in the range of moderate concentrations [Schuld & Wolf, 2001].

Ovejero et al. [Ovejero et al., 2009] determined Flory-Huggins parameter χ_{12}^{∞} for SEBS triblock copolymer. They noticed that Flory-Huggins parameter was defined as independent of concentration, but the effect of concentration is not negligible. Authors tried to develop a thermodynamic tool to simulate a polymer - solvent separation. In their work they also paid attention to temperature dependence. A decrease of Flory-Huggins parameter while increasing temperature was suggested. However they have shown that this dependence is not clear. For investigated rubber values of χ_{12}^{∞} increased slightly with temperature.

When mixture of components is used as a stationary phase in a chromatographic column, subscripts 2 and 3 are used to represent first and second mixtures' component, respectively [Voelkel et al., 2009 (b)]:

$$\chi_{1m}^{\infty} = \ln\left(\frac{273.15 \cdot R}{p_1^o \cdot V_g \cdot M_1}\right) - \frac{p_1^o}{R*T} \cdot \left(B_{11} - V_1^o\right) + \ln\left(\frac{\rho_1}{\rho_m}\right) - \left(1 - \frac{V_1^o}{V_2^o}\right) \cdot \varphi_2 - \left(1 - \frac{V_1^o}{V_3^o}\right) \cdot \varphi_3 \qquad (6)$$

where φ_2 and φ_3 are the volume fractions of components.

When $\chi < 0.5$, the probe liquid is generally characterized as a good solvent for the polymer, whereas $\chi > 0.5$ indicates a poor solvent which use may lead to phase separation. In the case of a polymer blend, the parameter χ can still be defined and the miscibility generally occurs when $\chi < 0$, because the high molar volume of both components diminishes the combinatorial entropy [Huang, 2009].

When a polymer blend is used the interaction between the two polymers is expressed in terms of χ'_{23} as an indicator of the miscibility of the components of the polymer blend. If the parameters χ_{12}^{∞} and χ_{13}^{∞} are known (from IGC experiment with appropriate component "2" or "3") the interaction parameter χ'_{23} may be calculated from equation [El-Hibri et al., 1989; Olabisi, 1975]:

$$\chi'_{23} = \frac{1}{\varphi_2 \cdot \varphi_3} \cdot (\chi_{12}^{\infty} \cdot \varphi_2 + \chi_{13}^{\infty} \cdot \varphi_3 - \chi_{1m}^{\infty}) \qquad (7)$$

Here, the second subscript of χ identifies the nature of the column.

The interaction between the two components of composition is expressed in terms of χ'_{23} may be also calculated from [Milczewska et al., 2001 as cited in Li (Pun Choi), 1996, Milczewska et al., 2003 as cited in Voelkel & Fall, 1997]:

$$\chi'_{23} = \frac{\chi^{\infty}_{23} \cdot V_1}{V_2} = \frac{1}{\varphi_2 \cdot \varphi_3} \cdot \left(\ln \frac{V_{g,m}}{W_2 \cdot v_2 + W_3 \cdot v_3} - \varphi_2 \cdot \ln \frac{V_{g,2}}{v_2} - \varphi_3 \cdot \ln \frac{V_{g,3}}{v_3} \right) \tag{8}$$

Here, the second subscript of V_g identifies the nature of the column. From Eq. (8), χ'_{23} may be calculated even for probes for which the parameters p^0_1, B_{11} and V^0_i are not known or are known with insufficient accuracy [Al-Saigh & Munk, 1984].

To obtain χ'_{23} for a polymer blend or composition utilizing IGC, χ^{∞}_{12} values for all components have to be known. Therefore, three columns are usually prepared: two for single components and the third one for a composition of the two components used. A further three columns containing different compositions of components can also be prepared if the effect of the weight fraction of the mixture on the examined property needs to be explored. These columns should be studied under identical conditions of column temperature, carrier gas flow rate, inlet pressure of the carrier gas, and with the same test solutes [Al-Saigh, 1997].

Large positive values of χ'_{23} indicates the absence or negligible interactions between components, a low value indicates favorable interactions, while negative value indicates strong interactions (the pair of polymers is miscible).

Equations (7) or (8) were frequently used to study the interaction parameter between two stationary phases using the IGC method. In literature data, it was found that, in many miscible systems, χ'_{23} values were probe dependent. The values of χ'_{23} were positive when χ^{∞}_{12} and χ^{∞}_{13} were positive, and decreased when χ^{∞}_{12} and χ^{∞}_{13} decreased to negative. Some negative χ'_{23} values were generally observed for probes with low χ^{∞}_{12} and χ^{∞}_{13} [Huang, 2009].

Nesterov and Lipatov [Nesterov & Lipatov, 1999] studied thermodynamics of interactions in the ternary system: polymer A + polymer B + filler S. In their studies it was shown that the introduction of a third component into the binary immiscible mixture of two polymers, where the third component is miscible with each component of binary mixture, may lead full miscibility of the ternary system. For the immiscible mixtures of polyolefins with polyacrylates and polymethacrylates it was discovered that a mineral filler (e.g. silica) also may serve as compatibilizer.

The compatibilization effect of two immiscible polymers by adding the third polymer (or filler) my be described in the framework of the Flory-Huggins theory extended for describing ternary mixtures. For that mixtures Flory-Huggins parameter can be expressed as:

$$\chi_{A+B+C} \cong \chi_{AB} \cdot \varphi_A \cdot \varphi_B + \chi_{AC} \cdot \varphi_A \cdot \varphi_C + \chi_{BC} \cdot \varphi_B \cdot \varphi_C \qquad (9)$$

A positive value of the parameter χ_{A+B+C} corresponds to an immiscible systems whereas a negative is an indicator of miscibility [Nesterov & Lipatov, 2001].

Values of Flory-Huggins χ'_{23} parameter depend on chemical structure of the solute and it is a common phenomenon, although not allowed by the theory [Fernandez-Sanchez et al., 1988]. It has been interpreted as a result of preferential interactions of the test solute with one of two components. This phenomenon for polymer blends was described independently by Fernandez-Sanchez et al. and Olabisi [Olabisi, 1975]. They attributed this to the non-random distribution of the solute in the stationary phase owing to its preferential affinity for one of the components. Selective solutes do not "sense" the three varieties of intramolecular contacts in the polymer mixture (A-A, A-B, B-B) in proportion to concentration. This modifies the retention volume (and χ'_{23}) values relative to those which would be obtained by truly random mixing of the solute with the polymer. Less selective solvents, on the other hand, exhibit a more random 'sampling' of the molecular environment of the stationary phase owing to the equal affinities they have for both. It is therefore expected that a better measure of the polymer-polymer interaction will be likely with less selective solvents.

Olabisi [Olabisi, 1975] described a polyblend as micro heterogeneous, where the size of the different phases and their interpenetration being limited by a host of factors among which are the extents of mixing, compatibility, molecular weight, clustering behaviour of each polymer, rheological and surface and interfacial properties. He attributed Flory-Huggins parameter dependence on test solute to unequal distribution of the solute in the stationary phase, and to wide range of interactions (polar, nonpolar, hydrogen-bonding and also electronic and electrostatic interactions) [Li, 1996]. Olabisi proposed to use a set of solvents based on their type of interactions with probe: (i) proton accepting strength, probed with chloroform and ethanol; (ii) proton donor strength with methyl-ethyl ketone and pyridine; (iii) polar strength with acetonitrile and fluorobenzene; (iv) nonpolar strength with hexane and carbon tetrachloride.

Prolongo et al. proposed to calculate the polymer-polymer interaction parameter χ from measurements performed on ternary systems composed of the polymer pair plus a solvent or probe. They given the expressions needed to calculate the true polymer-polymer χ based on the equation-of-state theory and they compared that method for PS+PVME data obtained from vapor sorption (VP). The results show that the values obtained from IGC correlation and VP are nearly the same [Prolongo et al., 1989].

Many authors suggest that the χ'_{23} values are solvent (solute) independent for probes giving $\chi_{12}^{\infty} = \chi_{13}^{\infty}$ [Su & Patterson, 1977; Lezcano et al., 1995]. If the difference between the interaction of the components (for blend $\equiv$ polymers 2 and 3) with the solvent is negligible $\left| \chi_{12}^{\infty} - \chi_{13}^{\infty} \right| = \Delta\chi \approx 0$ interaction parameter χ'_{23} should be solvent independent. The equation above is often called "$\Delta\chi$ effect".

Horta's group [Prolongo et al., 1989] have proposed a method based on the equation-of-state theory, which gives a polymer-polymer parameter χ_{23} named 'true', because the assumption that the Gibbs mixing function for the ternary polymer-polymer-solvent system is additive with respect to the binary contributions is avoided. They suggested that it is necessary to substitute the volume fraction φ_i in the Flory-Huggins theory by segment fractions ϕ_i according to:

$$\phi_i = \frac{w_i \cdot v_i^*}{\sum w_i \cdot v_i^*}$$ (10)

where v_i^* and w_i represent characteristic specific volume and the weight fraction of the ith component, respectively.

Shi and Shreiber [Shi & Shreiber, 1991] stated that the probe dependence of χ_{23}' is due to two major contributing factors. Firstly, the surface composition of a mixed stationary phase will rarely, if ever, correspond to the composition of the bulk. Thermodynamic requirements to minimize the surface free energy of the stationary phase will favor the preferential concentration, at the surface, of the component with the lower (lowest) surface free energy. Thus, the values of φ_2 and φ_3, as defined by the bulk composition of mixtures, are inapplicable to Eq. (7). Instead, a graphical method was proposed by Shi and Schreiber to evaluate the effective volume fraction and to correct the problem. Secondly, since χ_{12}^∞ and χ_{13}^∞ will not usually be equal, it follows that the volatile phase will partition preferentially to the component that has the lower pertinent χ_{1m}^∞ value. Thus, the partitioning must vary with each probe, inevitably affecting the χ_{23}' datum.

Deshpande and Farooque were the first to suggest the use of IGC for studying polymer blends [Deshpandee et al., 1974]. Starting form the Flory-Huggins expression for the change of the free enthalpy in mixing, which was extended to three-component systems, they proposed a method of analysis of IGC measurements on polymer blends which yielded the polymer-polymer interaction parameter χ_{23}'. They also observed probe dependency and tried to develop a method to evaluate probe-independent interaction [Farooque et al., 1992].

Milczewska and Voelkel [Milczewska & Voelkel, 2006] mentioned some of that methods of evaluating probe-independent interaction parameter. One of the solution may be procedure proposed by Zhao and Choi [Zhao & Choi, 2001; Zhao & Choi, 2002]. Authors proposed to use 'common reference volume' which vanishes the problem. As the reference volume they used molar volume of the smallest repeated unit of polymer.

Flory-Huggins parameter for blends can be calculated from equations:

$$\chi_{1m} = \frac{V_o}{V_1} \cdot \left(\ln \frac{273.15 \cdot R}{M_1 \cdot V_g \cdot p_1^o} - 1 + \left(1 - \frac{V_1}{V_2}\right) \cdot \varphi_2 + \left(1 - \frac{V_1}{V_3}\right) \cdot \varphi_3 - \left(\frac{B_{11} - V_1}{R \cdot T}\right) \cdot p_1^o \right)$$ (11)

and

$$\chi_{1m} = \varphi_2 \cdot \chi_{12} + \varphi_3 \cdot \chi_{13} - \varphi_2 \cdot \varphi_3 \cdot \chi_{23}'$$ (12)

Equation (12) predicts that a plot of χ_{1m} versus $(\varphi_2 \cdot \chi_{12} + \varphi_3 \cdot \chi_{13})$ will give a straight line with a slope 1 and an intercept of $-\varphi_2 \cdot \varphi_3 \cdot \chi_{23}'$.

Jan-Chan Huang [Huang, 2003] and with R. Deanin [Huang & Deanin, 2004] rearranged equation (12) into the following form:

$$\frac{\chi_{1m}}{V_1} = \frac{\varphi_2 \cdot \chi_{12} + \varphi_3 \cdot \chi_{13}}{V_1} - \frac{\varphi_2 \cdot \varphi_3 \cdot \chi_{23}'}{V_2} \qquad (13)$$

The polymer-polymer interaction term can be determined from the intercept at $\left(\dfrac{\varphi_2 \cdot \chi_{12} + \varphi_3 \cdot \chi_{13}}{V_1} \right) = 0$. This modification provided smaller standard deviations for the slope and the polymer-polymer interaction parameter.

Jan-Chan Huang [Huang, 2006] used also solubility parameter model to the study of the miscibility and thermodynamic properties of solutions by means of IGC. Because polymer–polymer mixtures have little entropy of mixing, the miscibility is largely decided by the sign of the heat of mixing. The determination of the heat of mixing becomes the key factor. The heat of vaporization is related to the solubility parameter, δ, of the liquid by the relation:

$$\delta = \left(\frac{\Delta E_{vap}}{V} \right)^{1/2} \qquad (14)$$

where ΔE_{vap} is the energy of vaporization and V is the molar volume of the solvent.

The Flory–Huggins interaction parameter can be related to the solubility parameters of the two components by:

$$\chi = \left(\frac{V_1}{RT} \right) \cdot (\delta_1 - \delta_2)^2 \qquad (15)$$

where δ_1 and δ_2 are the solubility parameters of the solvent and polymer, respectively, and V_1 is the volume of the solvent.

Guillet and co-workers [DiPaola-Baranyi & Guillet, 1978; Ito & Guillet 1979] have proposed IGC method for estimating of Flory–Huggins interaction parameter and solubility parameter for polymers by the modification of Eq. (15):

$$\left(\frac{\delta_1^2}{RT} - \frac{\chi}{V_1} \right) = \left(\frac{2\delta_2}{RT} \right) \delta_1 - \left(\frac{\delta_2^2}{RT} \right) \qquad (16)$$

It is a straight line equation. The left-hand side contains the values of Flory–Huggins interaction parameter of test solute (see Eq. (2)), solubility parameter of test solute (δ_1)) and its molar volume. Plotting the left-hand side of such equation vs. solubility parameter of test solute (δ_1) one obtains the slope ($a = 2\delta_2/RT$) enabling the calculation of the solubility parameter of the examined material. This value should be equal to that found from the intercept and positive [Voelkel et al., 2009 (b)].

When a mixture is used as the stationary phase the solubility parameter of the mixture, δ_m, can be compared with the prediction of the regular solution method, which gives δ_m to be the volume average of the two components [Huang, 2006]:

$$\delta_m = \varphi_A \cdot \delta_A + \varphi_B \cdot \delta_B \tag{17}$$

From this equation the formula of specific heat of mixing in the regular solution theory could be derived. A measurement of the solubility parameter of the polymer mixtures would then be a good indicator to predict their miscibility.

Huang [Huang, 2006] proposed a mechanism of probe dependency. When two polymers with specific interactions are brought together some functional groups interact with each other and are no longer available to the probes. Relative to the volume average of the pure components the probes will feel the mixture becomes lower in polar or hydrogen bonding interaction and more in nonpolar dispersive force. In other words, the mixture becomes more "alkane-like". The polar probes will be squeezed from the stationary phase and the specific retention volume decreased, which increases χ_{1m} through Eq. (5) then decreases

χ'_{23} through Eq. (13). Therefore, polar probes have lower retention volume and χ'_{23}, and for n-alkane probes the change is less. This difference between probes is exhibited as the probe dependency.

4. Applications

Authors examined many polymeric materials filled with modified silica or other inorganic fillers. Our measurements were carried out with the use of Chrom5 (Kovo, Prague, Czech.Rep.) gas chromatograph equipped with a flame ionisation detector. Some of the results were presented here.

4.1 Flory-Huggins parameters for polylactic acid compositions

For composition of polylactic acid (P, M=55000), containing different amount (5, 10, 15% wt) of modified silica (B2 and B5) [Jesionowski, 1999] or modified carbonate-silicate fillers (N1 and N2) [Grodzka 2004] we calculated Flory-Huggins parameters χ_{12}^{∞} and χ'_{23}. The influence of the temperature and the amount and type of filler was examined. To eliminate the solvent dependence of χ'_{23} values (from basic Eq. 8) experimental data were recalculated according to Zhao-Choi procedure.

Small volumes (0.5µL) of vapour of the probes were injected manually to achieve the infinite dilution conditions. These were: n-pentane (C5), n-hexane (C6), n-heptane (C7), n-octane (C8), n-nonane (C9), dichloromethane (CH2Cl2), chloroform (CHCl3), carbon tetrachloride (CCl4), 1,2-dichloroethane (Ethyl. Chl.) (all from POCH, Gliwice, Poland).

Values of χ_{12}^{∞} parameter for P-15N1 (it denotes the composition of polylactic acid with 15% of N1 filler) composition we presented in Figure 2. We obtained almost the same values of χ_{12}^{∞} parameter for other investigated compositions.

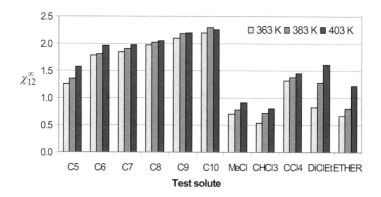

Fig. 2. Values of Flory-Huggins χ_{12}^{∞} parameter for P-15N1 composition

The lowest values of χ_{12}^{∞} parameter were obtained for dichloromethane (MeCl) and chloroform (CHCl3) as the test solute. Values calculated for compositions were almost always lower than those found for pure components, i.e. polymer and/or filler separately. The increase of temperature decreased values of χ_{12}^{∞} only for P-5B5 system, indicating the increase of interactions between composition and test solute. For the other compositions the increase of temperature increased values of χ_{12}^{∞} parameter.

The influence of the amount and type of filler was also examined. The change of these two factors also lead to the changes in the solute-composition interactions (Fig. 3). The influence of the amount of the filler is different for various compositions. For composition with N1 filler (carbonate-silicate filler modified with N-2-aminoethyl-3-aminopropyl-trimethoxysilane) the strongest interaction with solvent was found for the composition containing 5% of the filler. However, for P-N2 the most active is the composition with 15% addition of N2 (carbonate-silicate filler modified with n-octyltriethoxysilane).

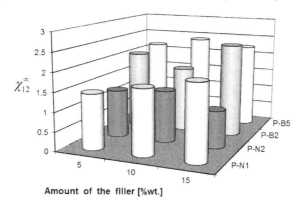

Fig. 3. The influence of the type and the amount of the filler [% wt.] on χ_{12}^{∞} parameters at 403K

Determined values of χ'_{23} depend on the type of the test solute used in IGC experiment (Fig. 4). Influence of the amount of the filler on the Flory-Huggins parameter χ'_{23} was examined and some results are presented in Figure 4. It is depended on test solute used in our study. For C5-C7 and CHCl3 and CCl4 we obtained the lowest values for 5% of N1 filler. For the other solutes – the strongest interaction are observed between polymer and 15% of the filler. Generally, the strongest interaction between components were observed for compositions with 5% of the filler.

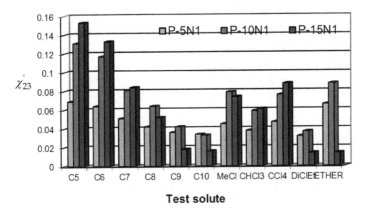

Fig. 4. Values of Flory-Huggins χ'_{23} parameter for compositions with 5%, 10% or 15% of filler

To eliminate the solvent dependence of χ'_{23} values (from basic equation) experimental data were recalculated according to Zhao-Choi procedure (Eq. 12). Values of $ZC\chi'_{23}$ parameter are presented in Figure 5. In all cases only one value for each composition was obtained. All $ZC\chi'_{23}$ values indicated the presence of strong or medium interaction between the modified filler and polymer matrix. This observation is consistent with that formulated after analysis of χ'_{23} data found for most of test solutes in the classic procedure.

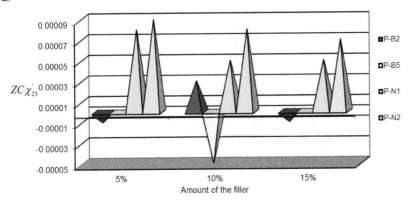

Fig. 5. Values of Flory-Huggins $ZC\chi'_{23}$ parameter for all compositions calculated according to Zhao-Choi procedure

It is worth to note that the increase of the filler content does not enhance the magnitude of interactions. Most often limited (rather negligible) decrease of polymer-filler interactions was observed.

4.2 Chemometric evaluation of IGC data

Principal Component Analysis (PCA) became a popular technique in data analysis for classification for pattern recognition and dimension reduction. It can reveal several underlying components, which explain the vast majority of variance in the data [Héberger, 1999; Malinowski, 1991; Héberger et al., 2001]. The principle is to characterize each object (rows in the input matrix) not by analyzing every variable (columns of the input matrix) but projecting the data in a much smaller subset of new variables (or principal component scores). PCA should facilitate the overcoming of the problem connected with the solute dependence of χ'_{23} parameter.

Values of Flory–Huggins χ'_{23} parameter expressing the magnitude of interactions between the polymer matrix and filler strongly depend on the type of test solute being used in IGC experiment (see Fig. 4). It causes the difficulties in the analysis of the influence of the type and amount of the filler onto the magnitude of these interactions. Such analysis is possible with the help of PCA technique [Voelkel et al., 2006] as presented in Fig. 6 for systems of polyurethane (PU) with modified silica fillers (B2). Materials used in experiments were described elsewhere [Milczewska & Voelkel 2002; Milczewska, 2001]. The magnitude of interactions is similar (the corresponding points belong to one – large cluster) for most of samples. Outside this large cluster the points correspond mainly to the compositions with 5 or 20% of the filler.

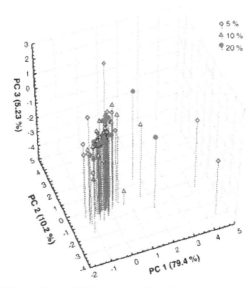

Fig. 6. Scatterplot for Polyurethane (PU) systems [Reprinted from Voelkel et al., 2006 with permission from Elsevier]

IGC procedures discussed earlier allow eliminating the test solutes dependence of χ'_{23} values. However, very often the relatively significant error of the determination was reported.

Values of χ'_{23} parameter calculated according to Zhao–Choi procedure for the examined polymeric composition are presented in Table 1. All values are negative and close to zero. It indicates the existence of polymer–filler interaction although their strength is limited.

Filler	PU-B2 compositions		
	PU+5%	PU+10%	PU+20%
IB2	$-2*10^{-5}$	$-2*10^{-5}$	$-2*10^{-5}$
IIB2	$-4*10^{-5}$	$-6*10^{-5}$	$-3*10^{-5}$
IIIB2	$-6*10^{-5}$	$-3*10^{-5}$	$-7*10^{-5}$
IVB2	$-3*10^{-5}$	$-1*10^{-5}$	$-7*10^{-5}$
VB2	$-7*10^{-5}$	$-3*10^{-5}$	$-1*10^{-5}$

Table 1. Values of χ'_{23} parameter calculated by Zhao-Choi method for B2-PU compositions [Reprinted from Voelkel et al., 2006 with permission from Elsevier]

The differences of the magnitude of polymer–filler interactions are significant as the error of determination is equal to approximately $2.5*10^{-7}$, i.e. it is at least two orders lower than the determined χ'_{23} values. However, collection of retention data for all test solutes is somewhat time-consuming. It would be useful to select the test solutes carrying the statistically valid information, applied these species in IGC experiments and further use their retention data in calculations of χ'_{23} from Zhao–Choi procedure. The problem was: how the reduction of the number of test solutes will influence the χ'_{23} values as well as error of their determination.

For all PU compositions PCA made possible using three - four test solutes (C6, C8, MeCl and CCl4) for determination of interaction parameters. Recalculating of χ'_{23} from Zhao–Choi procedure for selected test solutes gave values presented in Table 2.

Comparison of values of χ'_{23} calculated by Zhao-Choi method before and after PCA selection of solutes is presented on Figure 7. Corrected values are lower or higher than these found for all test solutes, but they indicate the presence or absence of interaction.

Filler	PU-B2 compositions		
	PU+5%	PU+10%	PU+20%
IB2	$-3*10^{-5}$	$-2*10^{-5}$	$-2*10^{-5}$
IIB2	$-3*10^{-5}$	$-5*10^{-5}$	$-3*10^{-5}$
IIIB2	$-5*10^{-5}$	$-3*10^{-5}$	$-5*10^{-5}$
IVB2	$-2*10^{-5}$	$0*10^{-5}$	$-7*10^{-5}$
VB2	$-4*10^{-5}$	$-1*10^{-5}$	$-1*10^{-5}$
Error ~ $2.5*10^{-7}$			

Table 2. Values of χ'_{23} parameter for B2-PU compositions calculated by Zhao-Choi method after PCA selection of solutes [Reprinted from Voelkel et al., 2006 with permission from Elsevier]

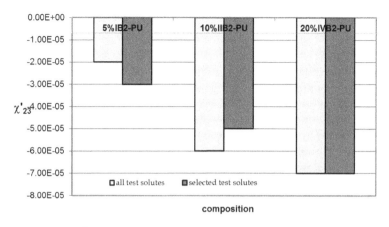

Fig. 7. Comparison of χ'_{23} calculated by Zhao–Choi procedure before and after PCA selection of test solutes for PU [Reprinted from Voelkel et al., 2006 with permission from Elsevier]

PCA enabled the significant reduction of the number of test solutes required for the proper determination of Flory–Huggins parameter and further the reduction of time required for proper characterization of examined material.

5. Summary

Inverse gas chromatography method has been found to be an effective tool for measurement of the magnitude of interaction (χ^{∞}_{12} interaction material – test solute) in polymers at different temperatures. From technological point of view χ_{23} is more interesting. This parameter can be used as indicator of the miscibility of the polymer blend.

Drawback of χ_{23} estimated by classical procedure is the test solute dependence. The procedure of its elimination proposed by Zhao-Choi seems to be most appropriate. It leads to realistic values of χ_{23} with low error of determination. The full procedure of χ'_{23} with the use of large series of the test solutes might be time-consuming. The use of chemometric analysis enabled the reduction of the number of the test solutes without loss of information. One may expect further application of IGC in examination of polymer materials.

6. References

Al-Ghamdi A., Al-Saigh Z.Y. (2000) *J. Polym. Sci., Part B: Polym. Phys.* Vol. 38 pp. 1155-1166.
Al-Saigh Z., Munk P. (1984) *Macromolecules* Vol. 17 pp. 803-809.
Al-Saigh Z.Y. (1997) *TRIP* Vol. 5 pp. 97-101.
Boukerma K., Piquemal J.Y., Chehimi M.M., Mravcakova M., Omastova M., Beaunier P. (2006) *Polymer* Vol. 47 pp. 569-576.
de Gennes P.G. (1979) *Scaling Concepts in Polymer Physics*, Cornell University Press, Ithaca, NY.

Deshpande D.D., Patterson D., Schreiber H.P., Su C.S. (1974) *Macromolecules* Vol. 7 pp. 530-535.

DiPaola-Baranyi G., Guillet JE. (1978) *Macromolecules* Vol. 11 pp. 228–235.

Dritsas G.S., Karatasos K., Panayiotou C. (2008) *J. Polym. Sci., Part B: Polym. Phys.* Vol. 46 pp. 2166-2172.

Dritsas G.S., Karatasos K., Panayiotou C. (2009) *J. Chromatogr. A* Vol. 1216 pp. 8979-8985

El-Hibri M.J., Cheng W., Hattam P., Munk P. (1989) Inverse Gas Chromatography of Polymer Blends. Theory and Practice, *in:* D.R. Lloyd, T.C. Ward, H.P. Schreiber (ed), *Inverse Gas Chromatography. Characterization of Polymers and Other Materials,* ACS Symposium Series, 391, Washington, 121-134.

Farooque A.M., Deshpande D.D. (1992) *Polymer* Vol. 33 pp. 5005-5018.

Fernandez M.L., Higgins J.S., Horst R., Wolf B.A. (1995) *Polymer* Vol. 36 pp. 149-154

Fernandez-Sanchez E., Fernandez-Torres A., Garcia-Dominguez J.A., Santiuste J.M., Pertierra-Rimada E. (1988) *J. Chromatogr.* Vol. 457 pp. 55-71.

Grobler A., Balizs, G. (1974) *J. Chromatogr. Sci.* Vol. 12 p. 57

Grodzka J. (2004) Ph.D. thesis, Poznan University of Technology, Poznan (in polish).

Gundert F., Wolf B. A. (1989) "Polymer-Solvent Interaction Parameters" in: *Polymer Handbook,* Ed. J. Brandrup and E. H. Immergut Wiley Interscience, New York, 3rd edn., pp. VII/173-182

Han S.J., Lohse D.J., Condo P.D., Sperling L.H. (1999) *J. Polym. Sci.: Part B: Polym. Phys.* Vol. 37 pp. 2835-2844.

Héberger K. (1999) *Chemometrics Intell. Lab. Syst.* Vol. 47 pp. 41-49.

Héberger K., Milczewska K., Voelkel A. (2001) *J. Chromatogr. Sci.* Vol. 39 pp. 375-384.

Hindawi J., Higgins J.S., Galambos A.F., Weiss R.A. (1990) *Macromolecules* Vol. 23 pp. 670-674.

Horst R., Wolf B.A. (1992) *Macromolecules* Vol. 25 pp. 5291-5296.

Huang J.-C. (2003) *J Appl. Pol. Sci.* Vol. 90 pp. 671-680.

Huang J.-C. (2009) *J. Appl. Polym. Sci.* Vol. 113 pp. 4085-4091.

Huang J.-C., Deanin R.D. (2004) *J Appl. Pol. Sci.* Vol. 91 pp. 146-156.

Ito K., Guillet J.E. (1979) *Macromolecules* Vol. 12 pp. 1163–1167.

James A.T., Martin A.J.P. (1952) *Biochem. J.* Vol. 50 pp. 679-690.

Jesionowski T. (1999) Ph.D. thesis, Poznan University of Technology, Poznan (in polish).

Lee H.S., Kim W.N., Burns C.M. (1997) *J. Appl. Polym. Sci.* Vol. 64 pp. 1301-1308.

Lezcano E.G., Prolongo M.G., Coll C.S. (1995) *Polymer* Vol. 36 pp. 565-573.

Li Bincai (Pun Choi) (1996) *Rubber Chem. Techn.* Vol. 69 pp. 347-376.

Malinowski E. R. (1991) *Factor Analysis in Chemistry,* 2nd ed., John Wiley and Sons, New York.

Mani S., Malone M.F., Winter H.H. (1992) *Macromolecules* Vol. 25 pp. 5671-5676.

Meoer H., Strobl G.R. (1987) *Macromolecules* Vol. 20 pp. 649-654.

Milczewska K. (2001) Ph.D. thesis, Poznan University of Technology, Poznan (in polish).

Milczewska K., Voelkel A. (2002) *J. Chromatogr. A* Vol. 969 pp. 255-259.

Milczewska K., Voelkel A. (2006) *J Polym Sci Part B: Polym Phys* Vol. 44 pp. 1853–1862

Milczewska K., Voelkel A., Jęczalik J. (2001) *Macromol. Symp.* Vol. 169 pp. 45-55.

Milczewska K., Voelkel A., Jęczalik J. (2003) *Macromol. Symp.*Vol. 194 pp. 305-311.

Mutelet F., Ekulu G., Rogalski M. (2002) *J. Chromatogr. A* Vol. 969 pp. 207-213.

Nesterov A.E. and Lipatov Y.S. (1999) *Polymer* Vol. 40 pp1347-1349

Nesterov A.E., Lipatov Y.S., Ignatova T.D. (2001) *Eur. Polym. J.* Vol. 37 pp. 281-285.

Olabisi O. (1975) *Macromolecules* Vol. 8 pp. 316-322.

Oliva V., Mrabet B., Neves M.I.B., Chehimi M.M., Benzarti K. (2002) *J. Chromatogr. A* Vol. 969 pp. 261-272.

Ovejero G., Pérez P., Romero M.D., Díaz I., Díez E. (2009) *Eur. Polym. J.* Vol.45 pp. 590-594.

Petri H.M., Schuld N., Wolf B.A. (1995) *Macromolecules* Vol. 28 pp. 4975-4980.

Prolongo M.G., Masegosa R.M., Horta A. (1989) *Macromolecules* Vol. 22 pp. 4346-4351.

Schuld N., Wolf B.A. (2001) *J. Polym. Sci.: Part B: Polym. Phys.* Vol. 39 pp. 651-662.

Schwann D., Frielinghaus H., Mortensen K., Almdal K. (1996) *Phys. Rev. Lett.* Vol. 77 pp. 3153-3156.

Shi Z.H., Schreiber H.P. (1991) *Macromolecules* Vol. 24 pp. 3522–3527.

Shillcock I.M., Price G.J. (2003) *Polymer* Vol. 44 pp. 1027-1034.

SMS-*i*GC brochure 2002

Su C.S., Patterson D. (1977) *Macromolecules* Vol. 10 pp. 708-710.

Sun H., Rhee K.B., Kitano T., Mah S.I. (1999) *J. Appl. Polym. Sci.* Vol. 73 pp. 2135-2142.

van Asten A., van Veenendaal N., Koster S. (2000) *J. Chromatogr. A* Vol. 888 pp. 175-196.

Voelkel A., Fall J. (1997) *Chromatographia* Vol. 44 pp. 197-204.

Voelkel A., Milczewska K., Héberger K. (2006) *Analytica Chimica Acta* Vol. 559 pp. 221-226.

Voelkel A., Strzemiecka B., Adamska K., Milczewska K. (2009) *J. Chromatogr. A* Vol. 1216 pp. 1551-1566 (b)

Voelkel A., Strzemiecka B., Adamska K., Milczewska K., Batko K. (2009) Surface and bulk characteristics of polymers by means of inverse gas chromatography *in* A. Nastasovic, S. Jovanovic (eds.) *Polymeric materials*, Research Signpost, pp. 71-102. (a)

Ying Q., Chu B., Wu G., Linliu K., Gao T., Nose T., Okada M. (1993). *Macromolecules* Vol. 26 pp. 5890-5896.

Zhao L., Choi P. (2001) *Polymer* Vol. 42 pp. 1075-1081.

Zhao L., Choi P. (2002) *Polymer* Vol. 43 pp. 6677-6681.

Zou Q.C., Zhang S.L., Wang S.M., Wu L.M. (2006) *J. Chromatogr. A* Vol. 1129 pp. 255-261.

10

Multidimensional Gas Chromatography – Time of Flight Mass Spectrometry of PAH in Smog Chamber Studies and in Smog Samples

Douglas Lane[1] and Ji Yi Lee[2]
[1]Environment Canada,
[2]Chosun University,
[1]Canada
[2]Republic of Korea

1. Introduction

Hans Falk (Falk et al., 1956; 1960) published the first papers demonstrating that polycyclic aromatic hydrocarbons underwent a transformation when exposed to ultraviolet light and oxidants. His experiments, in which concentrated spots of individual PAH were irradiated under UV light and exposed to ozone, suggested that PAH were oxidized to unknown products at a very slow rate of about 1 to 5% per hour. However, in 1977, it was shown (Lane & Katz, 1977) that PAH were, indeed, very sensitive to ozone by exposing monolayer distributions of PAH adsorbed on glass plates to ozone and irradiation which closely matched solar irradiation in the 2500A to 5000A region of the spectrum from a quartz lamp. Under lamp irradiation and 0.19 ppm O_3, the half lives of Benzo[a]pyrene, Benzo[k]fluoranthene and Benzo[b]fluoranthene were 0.58 h, 3.9 h, and 4.2 h respectively (Lane, 1975). They also were the first to describe how multi-layering of PAH on particles could influence the apparent reaction rate of the PAH in Falk's experiment, and demonstrated that a rapid surface oxidation reaction formed a "crust" of oxidized products and was then followed by a slower penetration reaction involving the ozone accessing the subsurface material by penetrating the surface oxidized material – much like passing through a crust of bread. In the 1980s the group at the University of California, lead by Roger Atkinson began smog chamber studies to determine the rate of decomposition of a wide variety of PAH and PAH-related compounds when exposed to various oxidants. Their work clearly indicated that the OH radical was the dominant oxidant in the atmosphere (Atkinson et al., 1984; Atkinson, 1988). Most subsequent smog chamber studies to investigate the decomposition of PAH have utilized the OH radical as the oxidant of choice. The first report of a variety of oxidized products of the reaction of Naphthalene with the OH radical in a smog chamber was published in 1994 (Lane & Tang, 1994). Numerous publications have followed, describing the products of the OH radical reactions with naphthalene (Atkinson & Arey, 1994; Bunce et al., 1997; Sasaki et al., 1997; Mihele et al., 2002; Wang et al., 2007; Lee & Lane, 2009; Nishino & Arey, 2009); with acenaphene (Sauret-Szczepanski & Lane, 2004); with Alkylated naphthalenes (Banceu et al., 2001; Wang et al., 2007) and with phenanthrene (Helmig & Harger, 1994; Esteve et al., 2003; Lee & Lane, 2010).

Gas Chromatographic methods, because of their resolution and separation power over other analytical methods have been used to investigate the myriad of products found. With the advent of multidimensional gas chromatography, (GCxGC), the resolution and separation capabilities of gas chromatographic methods took a quantum leap forward. The combination of GCxGC with Time of Flight Mass Spectrometry, (GCxGC-TOFMS) has resulted in the most powerful instrument for the determination of oxidation products in chemical reactions. Numerous papers demonstrating the power of GCxGC-TOFMS for the determination of the products of the OH radical reactions with various PAH in smog chamber studies have been published. (Lee & Lane, 2009; 2010; Lane & Lee, 2010). Lee and Lane then searched for, and found, many of the oxidation products found the smog chamber studies in smog samples collected in Seoul, Republic of Korea (Lane & Lee, 2010).

An important aspect of looking for secondary organic compounds produced in smog chamber studies, is to determine how the products partition themselves between the gaseous and particulate phases. This has been facilitated by the development of Annular Diffusion Denuders for organic compounds. Johnson and co-workers (Johnson et al., 1985; Lane et al., 1988; 1992) utilized 8-channel, multi-annular denuders, coated with a combination of SE-54 silicone oil and crushed Tenax®. Although these denuders were primarily designed for the collection of ambient air samples for the analysis of chlorinated aromatics, they were also shown to be suitable for the determination of the gas/particle partitioning of PAH. In 1995 Gundel (Gundel et al., 1995) introduced a single-channel denuder coated with finely ground XAD-4 resin for the analysis of PAH in tobacco smoke. In 1993 a collaboration between Gundel and Lane resulted in the first Integrated Organic Gas And Particle Sampler (IOGAPS) (Gundel & Lane 1995) which was manufactured by URG in Chapel Hill NC. The denuder in this instrument was an 8-channel, 30cm long denuder operating at a flow rate of 16.7 L/min. For application to smog chamber studies, a smaller 5-channel 20-cm long denuder was utilized (Mihele et al., 2002). Coupled with a filter pack comprising a quartz or Teflon coated glass fiber filter backed up with 2 sorbent impregnated filters (SIFs) (Gundel & Herring, 1998; Galarneau et al., 2006), the denuder-filterpack proved extremely efficient for the determination of the gas particle partitioning of the products during the course of the smog chamber reaction studies.

In this chapter, the application of GCxGC-TOFMS to the determination of the phase-partitioned products of the OH radical reactions of PAH will be discussed. The utility of thermal desorption of samples directly into the GCxGC-TOFMS for the analysis of smog samples from the Republic of Korea will also be described.

2. Experimental

2.1 Smog chamber

The smog chamber in our laboratory is a 10 m^3 cylindrical chamber, essentially a cylindrical Teflon bag which is surrounded by a bank of fluorescent black lights. These may be turned on in 12 stages giving 12 levels of UV intensity. For the reactions described in this paper, full illumination was used. A fan, the blades of which were also coated with Teflon, inside the chamber caused efficient air mixing in the chamber. Air is supplied to the chamber by an AADCO model 737 pure air generator. The air in the chamber was monitored using a Columbia Science Chemiluminescence monitor for the appearance of ozone, and, when

Multidimensional Gas Chromatography –
Time of Flight Mass Spectrometry of PAH in Smog Chamber Studies and in Smog Samples

217

ozone started to appear, it indicated that the NO, added to prevent the presence of ozone, had been used up. The reaction study was then terminated. NO was monitored by a Thermo Environmental Instruments INC. Model 42C NO-NO$_2$-NO$_x$ instrument.

The OH radical was generated in the chamber by the photochemical reaction of isopropyl nitrite in the presence of NO (Mihele et al., 2002). When the fluorescent lamps were turned on, the OH radical was produced to a steady state in just a few milliseconds (Bunce et al, 1997). When the lamps were turned off, the OH radical disappeared almost as quickly, effectively stopping the reactions and freezing the product mix. Samples could then be withdrawn from the chamber for analysis of the products. After the samples had been collected, the lamps were turned back on for a specified period of time after which the lamps were turned off and another sample was collected. This process continued until the NO was used up and ozone began to appear.

2.2 Annular diffusion denuder sample collection and analysis

A 5-channel, 30 cm long annular diffusion denuder coated with finely ground (to an average particle diameter of 0.7 μm) XAD-4 resin and filterpack were used to collect samples from the smog chamber to evaluate the gas/particle partitioning of the products produced during the reactions. During collection of the air samples, the gas phase products were adsorbed on the denuder while the particle phase material passed through the denuder and were trapped on the filter in the filter pack. The filter pack consisted of a Teflon coated glass fiber filter and two XAD-4 sorbent impregnated filters (Gundel & Herring, 1998; Galerneau et al., 2006) to trap any products which might volatilize from the particles collected on the filter and pass through the filter. The denuder and the filters were solvent extracted with hexane (Lane & Gundel, 1996; Gundel & Lane, 1999). The solutions were reduced in volume, internal standards were added and the samples were injected into the GC × GC-TOFMS.

2.3 Multidimensional Gas Chromatography-Time of Flight Mass Spectrometry (GCxGC-TOFMS)

A Pegasus IV GCxGC-TOFMS (LECO Instruments Inc, Dearborne, MI) was used for the chemical separation and analysis of the products. The first dimension column (for separation based on vapor pressure), was a DB5-MS capillary column (30 m × 0.25 mm i.d. × 0.25 μm film thickness) and the second dimension column (for separation in the polar dimension) was a DB17-MS column (1.1 m × 0.18 mm i.d. × 0.18 μm film thickness). The secondary oven temperature was offset +10 °C relative to the main oven and the modulator temperature was offset +20 °C relative to the main oven. The temperature modulator was a liquid nitrogen cooled dual jet configuration with a cool time of 1.90 s between heating and cooling cycles. The main oven was commenced at 60 °C for 3.0 minutes then programmed at 3 C° /min to 300 °C and then held isothermally for 5 minutes. The interface between the Gas Chromatograph and the mass spectrometer was held at 320 °C and the ion source was maintained at 225 °C. Mass scans were taken from 35 to 600 daltons at 200 full scans.s[-1]. LECO's ChromaTOF software v 3.32 was used for the control of the system and for the collection of and processing of the data. National Institute of Standards and Technology (NIST), Willey and in-house PAH mass spectral libraries were used for the identification of the analytes. When the mass spectrum of a compound in the sample agreed with a library reference with a match of greater than 800 out of 1000, a positive identification was indicated.

2.4 Thermal desorption Multidimensional Gas Chromatography–Time of Flight Mass Spectrometry (TD-GC × GC-TOFMS)

Being provided with only a portion of a 47 mm diameter glass fiber filter that had collected particulate matter over a 24 hour period of time (less than 24 m^3 of sampled air) sensitivity concerns arose. It was realized that to extract and reduce the sample to one mL would result in injecting less than 0.1 percent of the total sample whereas, if two 4 mm punches of the filter were thermally desorbed, about 6.5% of the entire sample or 65 times more than a single liquid injection could be introduced into the GC column in a single injection. We coupled a Gerstel Thermal desorption (TDS-G) system to the front end of the Pegasus and thermally desorbed the filter punches. The TDS was interfaced directly to the GC of the Pegasus by a transfer line that was maintained at 320°C. To desorb the filter punches, the thermal desorption tube was ramped from 20°C to 300°C at 25°C/min and held for 3 min under a flow of He (BIP grade, Linde Gas, Canada). The extracted analytes were trapped at -60 °C using liquid N_2 as coolant in a programmable temperature vaporizing inlet system (PTV-CIS, Gerstel Inc., Baltimore, MD). The inlet was operated in the solvent venting mode. The thermal desorption system was then warmed to 20 °C. While the CIS was heated to 320 °C at 12 C^o min^{-1}, the analytes were transferred to the first dimension GC capillary column. During the transfer of analytes the TD transfer line was fixed at 320 °C.

3. Results

3.1 Chromatographic results

Analyses of the smog chamber samples of naphthalene, acenaphthene and phenanthrene revealed many oxidized products and included hydroxy, quinone and many ring opening

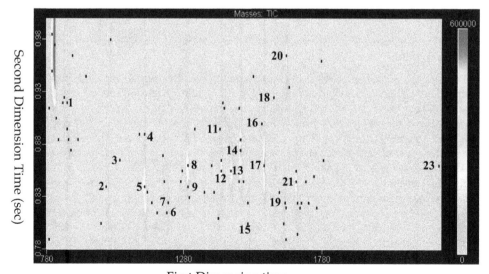

First Dimension time

Fig. 1. Showing the 2 dimensional plot of the compounds found in the gas phase and collected by the annular diffusion denuder.

Multidimensional Gas Chromatography –
Time of Flight Mass Spectrometry of PAH in Smog Chamber Studies and in Smog Samples

219

products (Sauret-Szczepanski & Lane, 2004; Lee & Lane, 2009; 2010). The results for the decomposition of naphthalene with the OH radical are shown in the contour chromatographic plots in Figures 1 and 2. In these figures, each black dot represents a distinctly separated and identified product. For some compounds in high concentration the colour surrounding the dot is an indication of intensity. Figure 1 shows the compounds that were trapped on the denuder and, therefore, were in the gas phase. Figure 2 shows the compounds that were extracted from the filter and SIFs. These were particle phase compounds. The numbers on the figures refer to the numbered compounds shown in Table 1 where the compound structure and the retention times on both columns are shown.

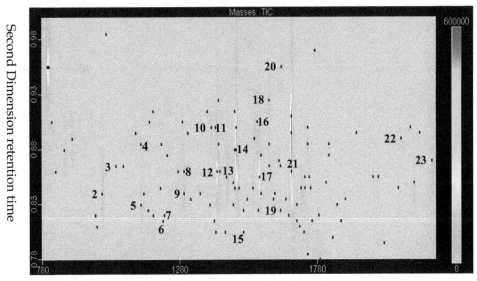

First Dimension retention time

Fig. 2. Showing the contour plot for the particle phase compounds found on the filter and the SIFs.

# in Fig. 1	Products	Structure	Formula	M.W.	1st R.T.*	2nd R.T*
1	1,4-dihydro-1,4-epoxy-Naphthalene					
2	Phthaladehyde		$C_8H_6O_2$	134	1000	0.85
3	Inden-1-one		C_9H_8O	132	1050	0.86

# in Fig. 1	Products	Structure	Formula	M.W.	1st R.T.*	2nd R.T*
4	1,2-benzopyrone		$C_9H_6O_2$	146	1140	0.895
5	Phthalic anhydride		$C_8H_4O_3$	148	1140	0.845
6	Phthalide		$C_8H_6O_2$	134	1220	0.815
7	1,3-indene-dione		$C_9H_6O_2$	146	1225	0.82
8	1,4-naphthoquinone		$C_{10}H_6O_2$	158	1295	0.865
9	1-hydroxy-naphthalen-2-one		$C_{10}H_8O_2$	160	1295	0.845
10	1-naphthalenol		$C_{10}H_8O$	144	1205	0.87
11	2-naphthalenol		$C_{10}H_8O$	144	1410	0.9
12	1-hydroxy-naphthalen-4-one		$C_{10}H_8O_2$	160	1415	0.86
13	2,3-epoxy-naphthoquinone		$C_{10}H_6O_3$	174	1450	0.855

Multidimensional Gas Chromatography –
Time of Flight Mass Spectrometry of PAH in Smog Chamber Studies and in Smog Samples

221

# in Fig. 1	Products	Structure	Formula	M.W.	1st R.T.*	2nd R.T*
14	(E)-2-formylcinnamaldehyde		$C_{10}H_8O_2$	160	1485	0.89
15	2,2-dihydroxy-indene-1,3-dione.		$C_9H_6O_4$	178	1510	0.805
16	1-nitronaphthalene		$C_{10}H_7NO_2$	173	1560	0.9
17	(Z)-2-formylcinnamaldehyde		$C_{10}H_8O_2$	160	1570	0.87
18	2-nitronaphthalene		$C_{10}H_7NO_2$	173	1605	0.92
19	1,2-naphthoquinone		$C_{10}H_6O_2$	158	1645	0.82
20	2-nitro-1-naphthol		$C_{10}H_7NO_3$	189	1650	0.965
21	2,3-epoxy-1-hydroxy-naphthalen-4-one		$C_{11}H_{12}O_2$	176	1685	0.84
22	4-nitro-1-naphthalenol		$C_{10}H_7NO_3$	189	2080	0.89
23	1-nitro-2-naphthalenol		$C_{10}H_7NO_3$	189	2195	0.865

Table 1. The products of the OH radical reaction of Naphthalene with the OH radical are shown together with the retention times on each column.

The products determined for Acenaphthene and Phenanthrene may be found in other publications (Sauret-Szczepanski & Lane, 2004; Lee & Lane 2010).

3.2 Challenges in relating TOFMS data to the NIST and Wiley databases

We soon discovered that TOFMS mass spectral data differ significantly from those in the NIST or WILEY databases that were largely derived from quadrupole mass spectrometric data. This is because of the manner in which the mass scans are obtained. With quadrupole mass spectrometers, maximum practical scanning rates are about 300 daltons.sec^{-1}, or about 1 to 1.5 full scans per second. This means that only about 3 scans can be taken across a single chromatographic peak. Because the mass of material being detected is constantly varying and because the scan takes about a second to be completed, the distribution of the peaks in a mass spectrum are biased high at the upper end of the spectrum as the peak is growing and are biased low at the upper end of the mass spectrum on the descending side of the peak. The reported mass spectra in the commercial databases are, of practical necessity, an average of the 3 or so peaks collected over one chromatographic peak. However, with TOFMS, full mass scans are taken at a rate of 200 per second. At such a rate, relative ion ratios are virtually constant at each point on a chromatographic peak. Approximately 600 spectra are taken over the width of a single chromatographic peak. It was for these reasons that chemical standards were obtained whenever possible and those standards used to generate an in-house library of TOFMS mass spectra. The agreement between samples and the in-house data were well above matches of 990 whereas the best matches with the NIST and WILEY libraries were on the order of 920-940. We had much greater confidence in the determinations of real world samples using our in-house library.

3.3 Retention times and mass spectral identification of products

Many of the products could be identified, although with lower match certainty than desired, through the use of the NIST and Wiley mass spectral databases, however, many more were not found in the databases and had to be determined by other means. For example, if standards or surrogate standards could be obtained mass spectra were obtained. It was also found useful to compare our spectra with mass spectral patterns published in the literature by other investigators. Their suggested identifications were of assistance in our own assignment of identities. Finally, when all else failed we identified the compounds through fundamental analysis of the mass spectra. To improve the match of environmental samples with the known products, we prepared an in-house database of the mass spectra of all reactants and for all products for which standards could be acquired. Thus matches between products in smog samples and database reference standards rose from about 650 to over 990 giving much greater confidence in the identity of the products.

3.4 Analysis of smog samples

A 3D image of the chromatograph of one of the Korean smog samples is shown below in Figure 3. From this sample, we successfully resolved almost 18,000 individual compounds. Many of the peaks could be identified by computer database searching. However, many were unidentified. This was partly due to the incompatibility of NIST and Wiley spectra with TOFMS data as outlined above and, more likely, because the compound in the air did not have a mass spectral signature in the databases.

Multidimensional Gas Chromatography –
Time of Flight Mass Spectrometry of PAH in Smog Chamber Studies and in Smog Samples

223

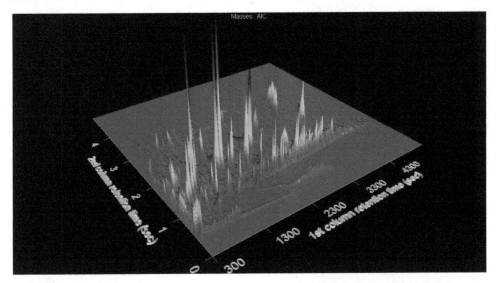

Fig. 3. The GCxGC trace of a Korean smog sample. The first column dimension is the vapor pressure dimension and is governed by chromatography on the DB-5MS column. The second dimension is the polar dimension and is governed by the polar DB-17MS column. The height of the peaks is proportional to the mass of the compound present.

However, for other unknowns or possibly improperly identified peaks, it was clearly a case of the compound not existing in the databases. We succeeded in finding 13 of the photochemical decomposition products from the chamber experiments in the Korean smog samples. The compounds from the decomposition of naphthalene, acenaphthene and phenanthrene that were found in the smog sample are shown in Table 2.

The above compounds were found in the particulate matter collected on the filter of a filter pack from Seoul, Republic of Korea. Since these compounds were found on the filter it is important to note that a) they were in the particle phase, 2) they may have suffered some volatilization or blow-off from the filter, but the degree to which that affected an estimation of the total compound in the atmosphere cannot be determined, and 3) many of the compounds detected such as the quinones, nitro derivatives and the hydroxynitro derivatives are known to be hazardous chemicals (Arey et al., 1989; Atkinson and Arey, 1994; Reisen and Arey, 2005). As many of these compounds, for example 1,2-naphthalenedicarboxaldehyde and (E)-2-formylcinnamaldehyde (Lane & Lee, 2010), are not known to originate in emissions, they must have been the result of atmospheric oxidation and this clearly indicates the formation of secondary aerosol material. It is important to note that there are no known anthropogenic sources of these two compounds. Recently Kroll and Seinfeld (Kroll and Seinfeld, 2008) have demonstrated the importance of atmospheric reaction products in the formation of secondary organic aerosol (SOA) and in the adverse effect of SOA to climate change and visibility in the atmosphere. A recent study (Robinson et al., 2007) has reported that research to estimate the organic aerosol budget, underestimates the production of SOA in the atmosphere when compared to actual field measurements. They suggested that the underestimate was due to the non-inclusion of the SOA produced from atmospheric semivolatile organic compounds.

Parent	Product	Structure
Naphthalene	1-naphthol	
	Phthalide	
	1,4-naphthalenedione	
	(E)-2-formylcinnamaldehyde	
	1,3-indandione	
	Phthalic anhydride	
	Indan-1-one	
Acenaphthene	1,8-naphthalicanhydride	
Phenanthrene	9-fluorenone	
	1,2-naphthalenedicarboxaldahyde	

Multidimensional Gas Chromatography –
Time of Flight Mass Spectrometry of PAH in Smog Chamber Studies and in Smog Samples

225

Parent	Product	Structure
	1-phenanthrol	
	9-phenanthrol	
	1,2-naphthalic anhydride	
	Dibenzopyranone	

Table 2. Above are presented the products found in the smog chamber during the studies of the reactions of naphthalene, acenaphthene and phenanthrene that were also found in the smog samples.

4. Conclusions

In this chapter, we have demonstrated that GC×GC-TOFMS is an excellent technique to identify the oxidized products of the decomposition of PAH in reactions with the OH radical. When combined with thermal desorption, TD-GC×GC-TOFMS becomes a very powerful tool to examine air extracts for a very wide range of pollutant chemicals. This method provides greatly enhanced sensitivity to chemical components and permits the detection of many, otherwise impossible to detect, compounds. We have demonstrated unequivocally that PAH are oxidized in the atmosphere and form a plethora of oxidized, nitrated and ring-opened products. This lends strong support to the statements of Robinson et al. (Robinson et al., 2007) that the production of SOA is underrepresented in budgets of atmospheric aerosols. We have found known oxidation products, some known only as atmospheric oxidation products with no anthropogenic source, in smog samples. This alone has many implications for the effect of SOA on human health.

5. References

Arey, J., Atkinson, R., Zlelinska, B., and McElroy, P.A. (1989). Diurnal concentrations of volatile polycyclic aromatic hydrocarbons and nitroarenes during a photochemical air pollution episode in Glendora, California. *Environmental Science & Technology.* 23, 321-327.

Atkinson, R., Aschmann, S.M., and Pitts, J.N. Jr. (1984). Kinetics of the reactions of Naphthalene and Biphenyl with OH radicals and with O_3 at 294 ± 1 K. *Environmental Science & Technology.* 18, 110-113.

Atkinson, R. (1988). Estimation of gas-phase hydroxyl radical rate constants for organic chemicals. *Environmental Toxicology & Chemistry.* 7, 435-442.

Atkinson, R., and Arey, J. (1994). Atmospheric chemistry of gas-phase polycyclic aromatic hydrocarbons: formation of atmospheric mutagens. *Environmental Health Perspectives.* 102, 117–126.

Banceu, C.E., Mihele, C.M., Lane, D.A., & Bunce, N.J. (2001). Reactions of Methylated Naphthalenes with Hydroxyl Radicals Under Simulated Atmospheric Conditions. *Polycyclic Aromatic Compounds.* 18, 415-425.

Bunce, N.J., Liu, L., Zhu, J., Lane, D.A., (1997). Reaction of naphthalene and its derivatives with hydroxyl radicals in the gas phase. *Environmental Science and Technology.* 31, 2252–2259.

Esteve, W., Budzinski, H., Villenave, E. (2003). Heterogeneous reactivity of OH radicals with phenanthrene. *Polycyclic Aromatic Compounds.* 23, 441-456.

Falk, H.L., Markul, I., and Kotin, P. (1956). Aromatic hydrocarbons. IV. Their fate following emission into the atmosphere and experimental exposure to washed air and synthetic smog. *A.M.A. Archives of Industrial Health* 13, 13-17.

Falk, H.L., Kotin, P., and Miller, A. (1960). Aromatic polycyclic hydrocarbons in polluted air as indicators of carcinogenic hazards. *International Journal of Air Pollution* 2, 201-209.

Galarneau, E., Harner, T., Shoeib, M., Kozma, M., and Lane, D.A. (2006). A preliminary investigation of sorbent-impregnated filters (SIFs) as an alternative to polyurethane foam (PUF) for sampling gas-phase semivolatile organic compounds in air. *Atmospheric Environment.* 40, 5734-5740.

Gundel, L.A., Lee, V.C., Mahanama, K.R.R., Stevens, R.K., and Daisey, J.M. (1995). Direct determination of the phase distributions of semi-volatile Polycyclic Aromatic Hydrocarbons using annular denuders. Atmospheric Environment. 29, 1719-1733.

Gundel, L.A., and Herring, S.V. (1998). Absorbing filter media for denuder-filter sampling of total organic carbon in airborne particles. Record of invention WIB 1457, Lawrence Berkeley National Laboratory (USA).

Gundel, L.A., and Lane, D.A. (1999). Sorbent-coated diffusion denuders for direct measurement of gas/particle partitioning by semi-volatile organic compounds.. In: Lane, D.A. (Ed.), Gas and Particle Phase Measurements of Atmospheric Organic Compounds. Gordon and Breach Science Publishers, Philadelphia PA, chapter 11.

Helmig, D., and Harger, W.P. (1994). OH radical-initiated gas-phase reaction products of phenanthrene. *Science of the Total Environment.* 148, 11-21.

Johnson, N.D., Barton, S.C., Thomas, G.H.S., Lane, D.A., and W.H. Schroeder. 1985. Development of a gas/particle fractionating sampler for chlorinated organics. In proceedings of the annual meeting of the Air Pollyution Control Association. Paper 85-81-1. June 12-21, Detroit, MI.

Kroll, J.H. and Seinfeld, J.H. (2008). Chemistry of secondary organic aerosol: formation and evolution of low-volatility organics in the atmosphere. *Atmospheric Environment.* 42, 3593-3624.

Multidimensional Gas Chromatography –
Time of Flight Mass Spectrometry of PAH in Smog Chamber Studies and in Smog Samples

227

Lane, D.A. (1975) Gas Chromatographic analysis and photodecomposition studies of atmospheric Polycyclic Aromatic Hydrocarbons. Ph. D. Thesis, York University, Toronto, Canada.

Lane, D.A., and Katz, M. (1977). The photodecomposition of benzo[a]pyrene, benzo[b]fluoranthene and benzo[k]fluoranthene under simulated atmospheric conditions. Advances in Environmental Science & Technology. 8, 137-154.

Lane, D.A., Johnson, N.D., Barton, S.C., Thomas, G.H.S., and W.H. Schroeder. (1988). Development and evaluation of a novel gas and particle sampler for semivolatile chlorinated organic compounds in ambient air. *Environmental Science & Technology.* 22, 941-947.

Lane, D.A., Johnson, N.D., Hanley, M.- J.J., W.H. Schroeder, and Ord, D.T. (1991). Gas- and particle-phase concentrations of α-hexachlorocyclohexane, γ- hexachlorocyclohexane, and hexachlorobenzene in Ontario air. *Environmental Science & Technology.* 26, 126-133.

Lane, D.A., Schroeder, W.H., and Johnson, N.D. (1992). On the spatial and temporal variations in atmospheric concentrations of hexachlorobenzene and hexachlorocyclohexane isomers at several locations in the province of Ontario, Canada. *Atmospheric Environment.* 26A, 31-42.

Lane, D.A. and Tang, H. (1994). Photochemical degradation of Polycyclic Aromatic Mihele, Compounds. I Naphthalene. *Polycyclic Aromatic Compounds.* 5, 131-138.

Lane, D.A., and Gundel, L.A. (1996). Gas and particle sampling of airborne polycyclic aromatic compounds. *Polycyclic Aromatic Compounds.* 9, 67-73.

Lane, D.A., Peters, A.J., Gundel, L.A., Jones, K.C., and Northcott, G.L. (2000). Gas/Particle partition measurements of PAH at Hazelrigg, UK. *Polycyclic Aromatic Compounds.* 20, 225-234.

Lane, D.A. and Lee, J.Y. (2010). Detection of known decomposition products of PAH in Particulate matter from Pollution Episodes in Seoul, Korea. *Polycyclic Aromatic Compounds.* 30, 309-320.

Lee, J.Y., and Lane, D.A. (2009). Unique products from the reaction of naphthalene with the hydroxyl radical. *Atmospheric Environment.* 43, 4886–4893.

Lee, J.Y., and Lane, D.A. (2010). Formation of oxidized products from the reaction of gaseous phenanthrene with the OH radical in a reaction chamber. *Atmospheric Environment.* 44, 2469-2477.

Mihele, C.M., Wiebe, H.A., Lane, D.A., (2002). Particle formation and gas/particle partition measurements of the products of the naphthalene-OH radical reaction in a smog chamber. *Polycyclic Aromatic Compounds.* 22, 729–736.

Nishino, N., Arey, J., and Atkinson, R. (2009). Formation and reactions of 2-formylcinnamaldehyde in the OH radical-initiated reaction of naphthalene. *Environmental Science & Technology.* 43, 1349-1353.

Reisen, F. and Arey, J. (2005). Atmospheric reactions influence seasonal PAH and nitro-PAH concentrations in the Los Angeles Basin. *Environmental Science & Technology* 39, 64-73.

Robinson, A.L., Donahue, N.M., Shrivastava, M.K., Weitkamp, E.A., Sage, A.M., Grieshop, A.P., Lane, T.E., Pierce, J.R. and Pandis, S.N. 2007. Rethinking organic aerosols: semivolatile emissions and photochemical aging. *Science.* 315, 1259-1262.

Sasaki, J., Aschmann, S.M., Kwok, E.S.C., Atkinson, R., & Arey, J. (1997). Products of the gas-phase OH and NO3 radical-initiated reactions of naphthalene. *Environmental Science and Technology*. 31, 3173–3179.

Sauret-Szczepanski, N., & Lane, D.A. (2004). Smog chamber study of acenaphthene: gas/particle partition measurements of the products formed by reaction with the OH radical. *Polycyclic Aromatic Compounds*. 24, 161–172.

Wang, L., Atkinson, R., and Arey, J. (2007). Dicarbonyl products of the OH radical-initiated reactions of naphthalene and the C1- and C2-alkylnaphthalenes. *Environmental Science & Technology*. 41, 2803-2810.

11

Recent Applications of Comprehensive Two-Dimensional Gas Chromatography to Environmental Matrices

Cardinaël Pascal[1], Bruchet Auguste[2] and Peulon-Agasse Valérie[1]
[1]SMS, Université de Rouen,
[2]CIRSEE (Centre International de Recherche Sur l'Eau et l'Environnement),
France

1. Introduction

Anthropological pressure on environment combined with continuous progress of analytical techniques allows the detection of more micro-pollutants in environmental matrices. Analysis of Persistent Organic Pollutants (POPs) remains a real challenge due to the large number of compounds and the complexity of environmental matrices. Conventional Gas Chromatography (GC) coupled with mass spectrometry (MS) is the reference technique for the analysis and the quantification of volatile and semi-volatile pollutants. Comprehensive two-dimensional gas chromatography (GC×GC) is a relatively new technique, developed in 1991 by Liu and Phillips (Liu & Phillips, 1991). This technique provides high separation power and sensitivity. The principles of multidimensional chromatography were described by Giddings (Giddings, 1984). When a fraction or few fractions of the effluent from a first column, is subsequently injected into a second column with a different selectivity, the multidimensional chromatographic separation techniques are called 'heart cutting'. These methods have proved to be very effective only in target compounds analysis. A two-dimensional separation can be called comprehensive if the three following conditions are established (Schoenmakers et al., 2003). First, every part of the sample is subjected to two different separations. Secondly, equal percentages (either 100% or lower) of all sample components pass through both columns and eventually reach the detector. Finally, the separation (resolution) obtained in the first dimension is essentially maintained. This latter point could be reached if the transfer of the effluent from the first column to the second one was successfully performed by a modulator or column interface. So, the modulator could be considered as the 'heart' of the system and is currently in development. A maximum of retention space could be used especially if compounds are subjected to two independent separations. Orthogonal separation occurs when the two columns use different separation mechanisms, operating independently in the two dimensions. In practice, columns containing chemically different stationary phases are chosen. In normal orthogonality, the first apolar column is coupled to a column containing a stationary phase of equivalent or higher polarity. For reversed orthogonality, the more polar stationary phase is used in first dimension and a less polar one in second dimension. Due to the low peak width, some constraints are imposed for the choice of detector. An ideal data acquisition rate for GC×GC detector is equal or more than 100 Hz to maintain its large separation power. Numerous detectors, conventionally used in GC like Flame Ionization Detector (FID), Electron Capture Detector (ECD) and microECD (μECD), have been widely employed. Concerning the MS, the high speed time-of-flight (TOFMS) with a unit-mass resolution has proved to be the best candidate for GC×GC. Moderate acquisition rate instruments, such as quadrupole mass spectrometer qMS (e.g. 20 Hz) were also used with a limited mass range. Several GC×GC instruments with various modulators have been developed (Semard et al., 2009) and are now commercially available. GC×GC has now demonstrated its capacity of resolution in the field of complex matrices like petroleum products, fragrance (Dallüge et al., 2003). GC×GC is currently one of the most effective techniques for the separation and analysis of environmental samples, offering significantly greater peak capacities than conventional chromatographic methods. GC×GC provides three major benefits, namely, enhanced chromatographic separation, improved sensitivity by effect of cryofocusing with the thermal modulator, and chemical class ordering in the contour plot (Ballesteros-Gomez & Rubio, 2011). Research in GC×GC has recently shifted from instrumental development to application to real samples over the last four years especially with the coupling to MS. Nearly 83% of the over 110 research articles published (specify the type or classification of articles) in 2010 were devoted to applications of GC×GC (Edwards et al., 2011). Moreover, some software improvements have also facilitated GC×GC quantitative analyses.

A few reviews were already published about applications of comprehensive GC in relation to environmental analyses. The most recent review was proposed by Wang et al. (Wang et al., 2010) that covered the works published between 2007 and the beginning of 2009. This chapter focuses on the most important developments in environmental applications of GC×GC, salient advances in GC×GC instrumentation and theoretical aspects reported over the period 2009 to July 2011. Recent applications using GC×GC methods for analysis of environmental toxicants such as PolyChlorinatedDibenzo-p-Dioxins (PCDDs), PolyChlorinatedDibenzoFurans(PCDFs), PolyChlorinatedBiphenyls (PCBs), Polycyclic Aromatic Hydrocarbons (PAHs), pesticides, alkylphenols *etc*… are reviewed. Moreover, this technique appeared especially suitable for the development of multiresidue analytical methods which was the most important trend in GC×GC-MS environmental analysis over the last period. The recent works demonstrated that this technique provides interesting alternative methods in terms of sensitivity and mapping of pollutants and minimizes sample preparation steps. A part of this chapter will be dedicated to the recent screening of emerging contaminants such as pharmaceuticals, plasticizers, personal care products, … Various matrices (water, soils and sediments) will be considered including river and wastewater. Analysis of air sample were not reported according to the publication of recent reviews dealing with the analysis of Volatile and semivolatile Organic Compounds (VOCs) found in the atmosphere (Arsene et al., 2011; Hamilton, 2010). All aspects of GC×GC will be presented including instrumentation, theoretical considerations and applications. Moreover, novel tools used for optimization of retention space or orthogonality estimation will be discussed.

2. GC×GC and environmental analysis reviews

A non exhaustive review (Ballesteros-Gomez & Rubio, 2011) was focused on main developments and advances in environmental analysis reported over the period 2009-2010.

Numerous aspects of environmental analysis, based on more than 200 articles, were reported including sampling, sample preparation, separation and detection ... Emerging contaminants and atomic spectrometry for the determination of trace metals and metalloids topics were excluded due to the publication of other reviews. Few applications of GC×GC were shortly reported (Eganhouse et al., 2009; Matamoros et al., 2010a; Hilton et al., 2010) and will be discussed in the present chapter. Recently, GC×GC applications devoted to measurement of volatile and semivolatile organic compounds in air and aerosol were reviewed and discussed (Arsene et al., 2011, Hamilton, 2010).

Wang *et al.* (Wang et al., 2010) have reviewed technological advances and applications of GC×GC between 2007 and July 2009. For example, separations of eight persistent organohalogenated classes of pollutants including OrganoChlorinatedPesticides (OCPs), PCBs, PolyBrominatedDiphenylEthers (PBDEs), PolyChlorinatedNaphthalenes (PCNs), PCDDs, PCDFs, PolyChlorinatedTerphenyls (PCTs), and toxaphene (CTT) in environmental samples were reported by Bordajandi *et al.* (Bordajandi et al., 2008). Nine column combinations in normal and reversed orthogonality (ZB-5, HT-8, DB-17 and BP-10, as first dimension column and HT-8, BPX-50 and Carbowax as second dimension one) were tested. The feasibility of the proposed approach for the fast screening of the target classes of pollutants was illustrated by the analysis of food and marine fat samples. A method of quantification of PAHs in air particulates based on a GC×GC isotope dilution mass spectrometry method was also reported by Amador-Munoz *et al.* (Amador-Munoz et al., 2009) with favorable resolution and sensitivity over conventional one dimensional GC. GC×GC-TOFMS method was successfully developed by Skoczynska *et al.* (Skoczynska et al., 2008) to identify 400 compounds in highly polluted sediment sample from the River Elbe (Czech Republic). Several older reviews dealing with the development of GC×GC and its applications including environmental matrices could be mentioned (Ramos et al., 2009; Cortes et al., 2009; Adahchour et al., 2006; Adahchour et al., 2008; Pani & Gorecki., 2006).

3. PCBs

PCBs are composed of 209 distinct congeners and are found in complex mixtures (Aroclors).They were commercially used in a variety of applications, including heat transfer and hydraulic fluids, dielectric fluids for capacitors, and as additives in pesticides, sealants, and plastics (Osemwengie & Sovocool, 2011). The World Health Organization (WHO) has designated twelve PCBs as "dioxin-like", coplanar PCB congeners that exhibited high toxicity.

Recently, a routine accredited method (Muscalu et al., 2011) was presented for analysis of PCBs, chlorobenzenes and other halogenated compounds in soil, sediment and sludge by GC×GC-µECD. A column combination DB1×Rtx-PCB was used to minimize coelution of analytes. The method was developed to analyze these pollutants in a single analytical run and no fractionation of sample extracts prior to instrument analysis, with enhanced selectivity and sensitivity over one dimensional GC method. The method can also be used to perform analytical triage to screen for additional compounds, for additional extract processing and testing or for identification and monitoring of new and emerging halogenated compounds present in sample extracts and to screen other halogenated organics. The optimized method provided quantification of Aroclors and Aroclors mixtures to within 15% of targets values and sub-nanograms per gram detection limits. The authors claimed that GC×GC requires minimal additional training to be used as a routine analytical method for the analysis of halogenated compounds.

Separation of 209 PCB congeners, using a sequence of 1D and 2D chromatographic modes was evaluated (Osemwengie et al., 2011). The authors used a RTX-PCB column as the first column and a DB-17 as the second one. In two consecutive chromatographic runs, 196 PCB congeners were distinguished, including 43 of the 46 pentachlorobiphenyl isomers. PCBs congeners that were not resolved chromatographically were resolved with the deconvolution program (ChromaTOFSoftware). Nevertheless, the 209 congeners have not been successfully separated.

New capillary columns coated with Ionic Liquids (ILs) were used as second columns for the separation of 209 PCBs congeners (Zapadlo et al., 2010; Zapadlo et al., 2011). In the first paper (Zapadlo et al., 2010), the orthogonality of three columns coupled in two series was studied. A non-polar capillary column coated with poly(5%-phenyl–95%-methyl)siloxane was used as the first column in both series. A polar capillary column coated with 70% cyanopropyl-polysilphenylene-siloxane or a capillary column coated with the ionic liquid 1,12-di(tripropylphosphonium)-dodecanebis(trifluoromethanesulfonyl)imide (IL36) was used as the second columns. The authors concluded that column coated with IL was more polar and more selective for the separation of PCBs than BPX-70 column (Figure 1).

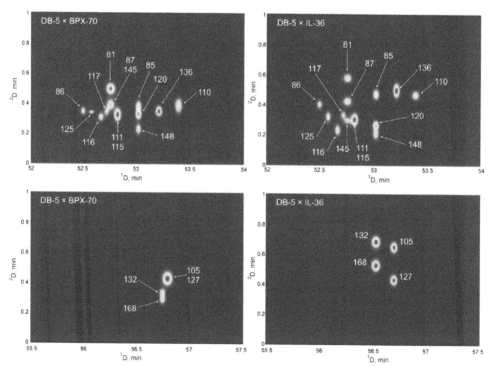

Fig. 1. 2D images for the separation of toxic, dioxin-like PCBs 81 and 105 on DB-5×BPX-70 and DB-5×IL-36 column series. Reprinted from Journal of Chromatography, A, (Zapadlo et al., 2010). Copyright (2010), with permission from Elsevier.

All "dioxin-like" PCBs, with the exception of PCB 118 and PCB 106, were resolved by this set of columns. In the second study (Zapadlo et al., 2011), the separation of 209 PCBs congeners was investigated using GC×GC–TOF-MS with a non-polar/IL column series consisting of poly(50%-n-octyl-50%-methyl)siloxane and (1,12-di(tripropylphosphonium)-dodecanebis(trifluoromethansulfonyl)amide) (SLB-IL59) in the first and second dimensions, respectively. A total of 196 out of 209 PCBs congeners were resolved by separation and/or mass spectral deconvolution using the ChromaTOF software. All "dioxin-like" congeners were separated with no interferences from any PCB congener. The 109 PCBs present in Aroclor 1242 and the 82 PCBs present in Aroclor 1260 were resolved on this column set.

A Quantitative Structure–Retention Relationship (QSRR) method (D'Archivio et al., 2011) was applied to predict the retention times of 209 PCBs in GC×GC. Predicted data were compared to GC×GC retention data taken from the literature. Authors demonstrated that the experimental GC×GC chromatogram of PCBs can be accurately predicted using a QSRR model calibrated with retention data of about 1/3 of the congeners collected under the same separation conditions. The effect of structure on retention time in both dimensions can be successfully encoded by theoretical molecular descriptors quickly available by means of various computational methods.

4. PCDDs and PCDFs

PCDDs and PCDFs constitute two classes of structurally related chlorinated aromatic hydrocarbons that are both highly toxic and produced as by-products during a variety of chemical and combustion processes. Due to their hydrophobic character and resistance to metabolic degradation, these substances exist as complex congener mixtures in the environment and are considered as POPs.

De Vos *et al.* (de Vos et al., 2011a) developed an alternative method of GC coupled with High Resolution Mass Spectrometry (HRMS) for analysis of PCDDs and PCDFs using GC×GC-TOFMS in different matrices. Three GC column combinations (Rtx-Dioxin 2×Rtx-PCB, Rxi-5 SilMS×Rtx-200 and Rxi-XLB×Rtx-200) were evaluated to quantify PCDDs and PCDFs in numerous soil and sediment samples taken from various strategic sites in South Africa with a highest result obtained of 76 ng Toxic Equivalent Quantity/kg. Results were also compared with those obtained using GC–HRMS and a good agreement was observed. The limit of detection (LOD) for the method (300 fg on column for spiked soil samples) was determined using the combination Rxi-XLB×Rtx-200 which provided excellent separation of the compounds mandated for analysis by United States Environmental Protection Agency (US EPA) Method (Figure 2).

Using a multi-step temperature program, all seventeen PCDDs and PCDFs components mandated by EPA Method 1613 were separated. GC×GC–TOFMS appeared to be a viable tool for dioxin screening and quantitation, especially in cases where PCDDs/PCDFs levels are greater than 1 ng.kg^{-1}. The technique proved to be ideal for application in developing countries where GC–HRMS is not available, and can be used to minimize costs by selecting only positive samples for further analysis by GC–HRMS. GC×GC–TOFMS additionally provides full range mass spectra for all sample components, thus allowing for identification of non-target analytes e.g. the brominated dioxins.

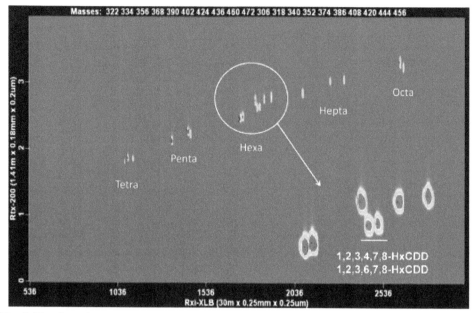

Fig. 2. 2D selected ion contour plot for the 17 priority PCDD/Fs using the Rxi-XLB/Rtx-200 column combination. The PCDD/Fs are again well resolved, especially the 1,2,3,4,7,8-HxCDD and 1,2,3,6,7,8-HxCDD isomers. Reprinted from Journal of Chromatography, A, (de Vos, 2011a). Copyright (2011), with permission from Elsevier.

GC×GC–TOFMS was also applied to investigate (de Vos et al., 2011b) toxic waste. This technique has allowed both comprehensive screening of samples obtained from a hazardous waste treatment facility for numerous classes of POPs and also quantitative analysis for the individual compounds. Various column combinations have been investigated for handling very complex waste samples. The close correlation between values obtained using the GC×GC-TOFMS approach and the GC-HRMS method has confirmed the validity of this technique to quantify PCDDs, PCDFs and four dioxin-like non-ortho substituted PCBs at levels required by regulatory bodies. The authors have obtained consistently higher values with the GC×GC–TOFMS method than those obtained with GC-HRMS. Nevertheless, they considered that the differences were certainly within permissible levels considering that the analyses have been performed in different laboratories.

An original application of the high sensitivity obtained using cryogenic zone compression (CZC) has been described by Patterson *et al.* (Patterson et al., 2011). The use of a GC×GC cryogenic loop modulator to perform CZC-GC coupled with Isotopic Dilution (ID) HRMS has been shown to be the most sensitive method available for the measurement of 2,3,7,8-tetrachlorodibenzo-p-dioxin (2,3,7,8-TCDD) (less than approximately 586,000 2,3,7,8-TCDD molecules) in human samples.

5. PAHs and hydrocarbons

PAHs are organic pollutants generated during the incomplete combustion of different natural and anthropogenic sources. They could enter the environment via municipal/industrial effluents. Exposure to PAHs represents a risk for human health due to their genotoxic and carcinogenic effects. The International Agency for Research on Cancer has classified them as possible and probable carcinogens for humans. The US EPA has included sixteen of them in the list of priority pollutants and establishes a maximum contaminant level for benzo[a]pyrene in drinking water at 0.2 μg.L^{-1}. In the European Union (EU), eight PAHs have been identified as priority hazardous substances in the field of water policy.

Chlorinated or brominated PAHs (Cl-PAHs and Br-PAHs) have been already detected in environmental samples such as fly ash (Horii et al., 2008) and sediment (Ishaq et al., 2003; Horii et al. 2009). Moreover, toxicities of Cl-PAHs have been investigated and reported (Horii et al., 2009). A method (Ieda et al. 2011) using GC×GC coupled with HRTOFMS was developed for the analysis of Cl-PAHs and Br-PAHs congeners in environmental samples. The GC×GC-HRTOFMS method allowed highly selective group type analysis with a very narrow mass window (e.g. 0.02 Da), accurate mass measurements for the full mass range (m/z 35–600) in GC×GC mode, and the calculation of the elemental composition for the detected congeners in the real-world sample. The authors reported, for the first time, the detection of highly chlorinated PAHs, such as $C_{14}H_3Cl_7$ and $C_{16}H_3Cl_7$, and ClBr-PAHs, such as $C_{14}H_7Cl_2Br$ and $C_{16}H_8ClBr$ in the environmental samples (Figure 3).

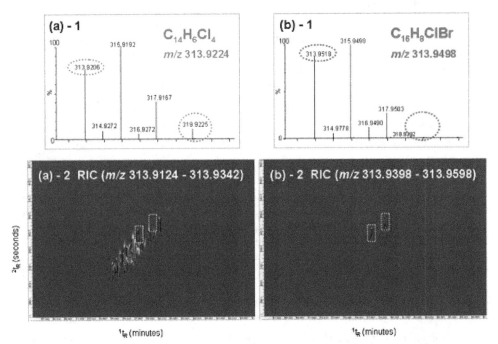

Fig. 3. The difference of isotope patterns between two peaks in the soil extract; (a)-1 $C_{14}H_6Cl_4$ and (b)-1 $C_{16}H_8ClBr$ and GC×GC–HR TOFMS 2D exact mass chromatogram of a 0.02 Da wide windows (a)-2 $C_{14}H_6Cl_4$; m/z 337.9224 and (b)-2 $C_{16}H_8ClBr$; m/z 313.9498. Reprinted from Publication, Journal of Chromatography, A, (Ieda et al., 2011). Copyright (2011), with permission from Elsevier.

Other organohalogen compounds; e.g. PCBs, PCNs, and PCDFs were also detected. This technique provided exhaustive analysis and powerful identification for the unknown and unconfirmed Cl-/Br-PAH congeners in environmental samples.

GC×GC-FID and GC×GC-TOFMS methods (Wardlaw et al., 2011) were used to study the biodegradation of alkylated naphthalenes and benzothiophenes isomers in marine sediment contaminated with crude oil. Their power resolution enabled separation and quantification of multiple structural isomers to determine their first order rate constants for aerobic biodegradation. Rate constants were used as proxies for microbial preference. A strong isomeric biodegradation preference was noted within each of these compound classes, with rate constants varying as much as a factor of 2 for structural isomers of the same compound class.

HPLC-GC×GC-FID and GC×GC-TOFMS were used to study the biodegradation of petroleum hydrocarbons in soil microcosms during 20 weeks (Mao et al., 2009). Aromatic hydrocarbons and n-alkanes were better biodegradable (>60% degraded) than iso-alkanes and cycloalkanes (<40%). GC×GC chromatograms showed that more polar and heavier compounds were formed as biodegradation proceeded.

6. Alkyl phenol isomers

AlkylPhenol EthOxylates (APEOs) are surfactants that have been widely used as detergents, emulsifier and dispersing agents in industrial or household cleaning products including laundry detergents. These compounds are degraded in wastewater treatment plants (WWTPs) in more toxic compounds, such as nonylphenols (NPs) and octylphenols (OPs). NPs and OPs, used for industrial production of APEOs surfactants, are complex mixtures of C_{3-10}-phenols where the main isomers are para-substituted. The interest in NPs and OPs analysis has increased during the last decades due to their capacity to disrupt the endocrine system which varies according to the structure of the branched alkyl group. They have been included in the water framework directive (WFD) as priority hazardous substances.

GC×GC was applied by Eganhouse *et al.* (Eganhouse et al., 2009) to enhance the chromatographic resolution of highly similar compounds such as 4-nonylphenol isomers and facilitate identification of a number of previously unrecognized components. Among the 153-204 peaks attributed to alkylphenol, 59-664-NPs were identified (Figure 4). Seven technical NPs products were analyzed using eight synthetic 4-NP isomers, with significant differences among the products and between two samples from a single supplier. This technique was also applied to environmental samples (wastewater, contaminated groundwater and municipal wastewater). The authors demonstrated that alteration of NPs composition through degradation results in enrichment of the more persistent isomers and removal or reduction of less persistent isomers. So, the estrogenicity may be increased or decreased depending on which 4-NP isomers are removed most rapidly.

The optimization of the separation of complex NPs technical mixtures (Vallejo et al., 2011) has been performed by means of experimental designs using GC×GC–FID and GC×GC–qMS equipped with valve-based modulator. Up to 79 OPs and NPs isomers have been separated using the FID detector and 39 have been undoubtedly identified using the mass spectra obtained from the qMS detector. The 22 OP, 33 OP, 363 NP and 22 NP isomers have been synthesized and quantified in two different technical mixtures from Fluka and Aldrich.

The values obtained for NP isomers were in good agreement with the literature and the values calculated for OP were for the first time reported.

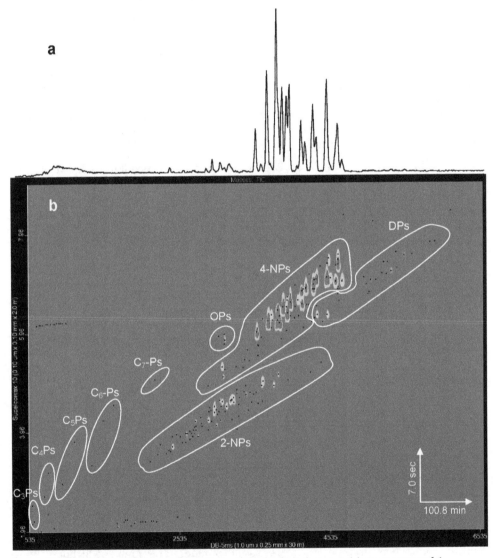

Fig. 4. Total ion chromatograms of technical NP (Fluka) showing (a) reconstructed 1-dimensional plot, and (b) 2-dimensional plot with alkylphenol regions indicated. Cx) C_xH_{2x+1}, OPs) octylphenols, DPs) decylphenols. x-axis represents the separation in the column with the nonpolar stationary phase (DB-5 ms), whereas the y-axis represents the separation in the column with the polar stationary phase (Supelcowax 10); retention time is given in seconds. Reprinted with permission from Environmental Science & Technology (Eganhouse et al., 2009). Copyright 2011 American Chemical Society.

7. Pesticides

Recently, a study (Macedo da Silva et al., 2011) demonstrated the potential of the application of GC×GC-μECD to the analysis of seven pesticide residues (propanil, fipronil, propiconazole, trifloxystrobin, permethrin, difenoconazole and azoxystrobin) in sediments. GC×GC-ECD method improved the separation between analytes and matrix interferences, minimizing the possibility of co-elutions. Its resolution capacity allowed the use of a selective detector instead of the use of a more expensive mass spectrometry detector. Best results were obtained with the set of columns DB-5×DB-17ms. The LODs for GC×GC method were about 36% lower than those obtained for the one dimensional GC method (in the range from 0.08 to 1.07 g.L^{-1}). Accuracy also indicated better results for GC×GC, possibly due to its higher sensitivity and lower contribution of co-eluting matrix components, which was minimized by increased peak capacity.

A GC×GC-qMS method (Purcaro et al., 2011) was developed for the multiresidue analysis of 28 pesticides contained in water. Pesticides extraction was performed by using direct Solid-Phase MicroExtraction (SPME). The rapid-scanning (20 000 amu/s) qMS system was operated using a rather wide m/z 50–450 mass range and a 33 Hz spectral production rate. The qMS performances were evaluated in terms of number of data points per peak, mass spectral quality, extent of peak skewing, and consistency of retention times.

A method for the determination of ultra-trace amounts of OCPs in river water was developed by Ochiai et al. using GC×GC–HR TOFMS (Ochiai & Sasamoto, 2011). Stir Bar Sorptive Extraction (SBSE) followed by thermal desorption (TD) was used for sample preparation. SBSE conditions including extraction time profiles, phase ratio (sample volume/PolyDiMethylSiloxane (PDMS) volume), and modifier addition were studied. The SBSE–TD–GC×GC–HR TOFMS method was solvent-free and highly selective and sensitive (LOD: 10–44 pg.L^{-1}). The method was successfully applied to the determination of 16 OCPs in river water sample. Authors showed that the results for 8 OCPs were in good accordance with the values obtained by a conventional Liquid–Liquid Extraction (LLE)–GC–HRMS (Selected Ion Monitoring) SIM method. The method also allowed the identification of 20 non-target compounds, e.g. pesticides and their degradation products, PAHs, PCBs and pharmaceuticals and personal care products and metabolites in the same river water sample, by using full spectrum acquisition.

A review dedicated to determination of pyrethroid insecticides in environmental samples was recently published by Feo et al. (Feo et al., 2010). The authors discussed the advantages and the disadvantages of the different instrumental techniques including GC×GC.

8. VOCs and other compounds

Benzothiazoles, benzotriazoles and benzosulfonamides are high-production-volume chemicals that are used in industrial and household applications. These compounds were detected in various environmental aqueous samples and were usually quantified by LC-MS/MS. Jover et al. (Jover et al., 2009) developed a Solid Phase Extraction (SPE)-GC×GC-TOFMS method for the characterization of benzothiazoles, benzotriazoles and benzosulfonamides in aqueous matrices. Columns combination was optimized to ensure a good separation between target analytes and interfering compounds of the matrix. 12 target analytes were characterized in river water and in wastewater from both the influent and the effluent of a WWTP. Similar method (Matamoros et al., 2010b) was used to study the benzothiazoles and benzotriazoles removal efficiencies of four WWTPs.

The methods for the determination of polycyclic and nitro-aromatic musk compounds as well as those for the respective metabolites are reviewed by Bester (Bester, 2009).The power of GC×GC approaches was demonstrated considering the various production impurities (isomers) of the two polycyclic musks with the highest usage rates.

A methodology to characterize VOCs and semi-volatile compounds from marine salt using HeadSpace (HS)-SPME and GC×GC-TOFMS was developed by Silva *et al.* (Silva et al., 2010). 157 VOCs distributed over the chemical groups of hydrocarbons, aldehydes, esters, furans, haloalkanes, ketones, ethers, alcohols, terpenoids, C_{13} norisoprenoids, and lactones were detected. Furans, haloalkanes and ethers were identified for the first time in marine salt. Contour plot analysis revealed the complexity of marine salt volatile composition and confirmed the importance of a high resolution, sensitive analytical procedure (GC×GC-TOFMS) for this type of analysis. The structured 2D chromatographic profile arising from ^{1}D volatility and ^{2}D polarity was demonstrated, allowing more reliable identifications. Results obtained for analysis of salt from two diverse locations and harvests over three years have suggested loss of volatile compounds according to storage duration of the salt, with environmental factors surrounding the saltpans influencing the volatile composition of the salt. At present the relative contributions of these factors have not been quantified. Origins of newly identified compounds in marine salt were in accordance with previous propositions, with algae, surrounding bacterial community, and environmental pollution being obvious sources.

9. Screening

In environmental monitoring, pollutants lists are periodically updated by regulatory agencies. Both the European Union (EU) and US EPA issued dangerous and hazardous contaminant lists, the so-called priority substances, whose concentration and occurrence in waters were strictly regulated (Directive 2000/60/EC; Decision No.2455/2001/EC and Clean Water Act) (Matamoros et al., 2010a). As the number of environmental regulated pollutants increases, it is necessary to develop global detection methods which can be used to screen a large number of substances simultaneously. This kind of method could reduce cost and time necessary for their detection and quantification. Moreover, there is a diverse group of unregulated pollutants called "emerging" contaminants, including pharmaceuticals and personal care products which were interesting to identify and monitor due to their high mass discharge into the environment. Some emerging contaminants have been recently included in candidate contaminant lists either from US EPA and the EU commission.

Semard *et al.* (Semard et al., 2008a; Semard et al., 2008b) reported a GC×GC-TOFMS method to search 58 target compounds and screen hazardous contaminants including PBDEs, PAHs and pesticides in urban wastewater. A variety of drugs (antidepressants, antibiotics, anticoagulants, *etc...*), personal care products (sunscreens, antiseptics, cosmetics etc.) and carcinogenic compounds, pesticides and compounds toxic for reproduction were identified in the raw wastewater. Most of these compounds were removed or decreased by the WWTP. Four priority substances (1,2,3-trichlorobenzene, 4-tert-butylphenol, benzothiazole and naphthalene) were present with concentrations in the range of 0.05 to 1.5 mg.L^{-1} in the raw wastewater and 0.01 to 0.1 mg.L^{-1} in the treated wastewater.

Household dusts were investigated using GC×GC-TOFMS as efficient screening method (Hilton et al., 2010). PAHs, phthalates, and compounds containing chlorine, bromine, or nitro groups were located on the chromatogram. Household dust (SRM-2585) was extracted with hexane using accelerated solvent extraction (ASE). Large molecules, such as triglycerides and fatty acids were removed with gel permeation chromatography. The resulting peak table was automatically filtered to identify compound classes such as phthalates, PAHs (Figure 5) and their heterocyclic analogs, PCBs, PBDEs, chloroalkyl phosphates, pesticides, and pesticides degradation products... By comparison with concentrations determined by National Institute of Standards and Technology, the technique was able to identify analytes at concentrations as low as 10–20 ng.g^{-1} dust for compounds quantified by NIST (National Institute of Standards and Technology).

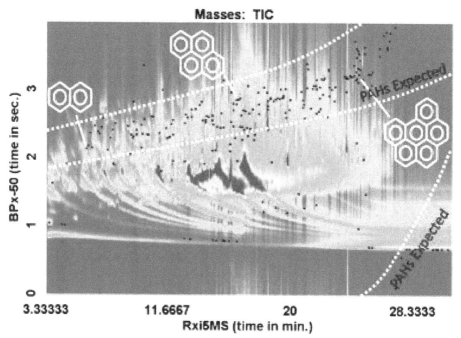

Fig. 5. Location of peaks matching the PAH spectral pattern with those identified by library searching. PAHs can be expected to fall into a band, shown starting at about 2 seconds in the second dimension and, as the chromatogram proceeds, falling later in the second dimension. Reprinted from Journal of Chromatography, A, (Hilton et al., 2010). Copyright (2010), with permission from Elsevier.

A SPE-GC×GC-TOFMS screening method for 97 priority and emerging contaminants in river was developed by Matamoros et al. (Matamoros et al., 2010a). The SPE was followed by in GC-port methylation using trimethylsulfonium hydroxide. The target analytes included 13 pharmaceuticals, 18 plasticizers, 8 personal care products, 9 acid herbicides, 8 triazines, 10 organophosphorous compounds, 5 phenylureas, 12 organochlorine biocides, 9

PAHs, 5 benzothiazoles and benzotriazoles. Best resolution between matrix constituents and target analytes was observed with TRB-5MS×TRB-50HT (apolar – polar) columns combination. Moreover, using polar-nonpolar columns combination, a strong correlation between the second dimension retention time and log K_{ow} for the target compounds was observed and was proposed as an additional identification criterion. The method was successfully applied to the analysis of four river water samples with LOD ranging from 0.5 to 100 ng.L[-1] (Figure 6). Plasticizers (e.g., phthalates and bisphenol A), pharmaceuticals (e.g., naproxen, ibuprofen), and personal care products (e.g., tonalide and methyl dihydrojasmonate) were the most abundant in concentration and detection frequency.

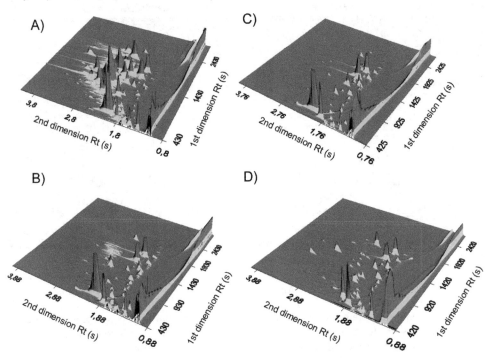

Fig. 6. 3D contour plots of four rivers sampled in this study in which the total ion chromatogram is shown: Ebro (A), Llobregat (B), Ter (C), and Beso`s (D). Reprinted with permission from Analytical Chemistry, (Matamoros et al., 2010a). Copyright 2011 American Chemical Society.

Gomez et al. (Gomez et al., 2011) developed a GC×GC–TOFMS method for the automatic searching and evaluation of nonpolar or semipolar contaminants (13 personal care products, 15 PAHs and 27 pesticides) in wastewater and river water. SBSE was selected for sample preparation step. Good results have been obtained in terms of separation efficiency and detection limits at or below 1 ng.L[-1] for most of the compounds in the MS full scan mode, using only 100 mL of river water sample and 25 mL of wastewater effluent sample. The authors mentioned the possibility to screen for non-target compounds or unknowns. New contaminants have been identified in the wastewater effluents and river water samples, such as cholesterol and its degradation products, pharmaceuticals, illegal drugs, industrial products as well as other pesticides and personal care products. Moreover, GC×GC features were proposed to compare the fingerprinting of different water samples giving valuable information about the contamination status of rivers and wastewaters (Figure 7).

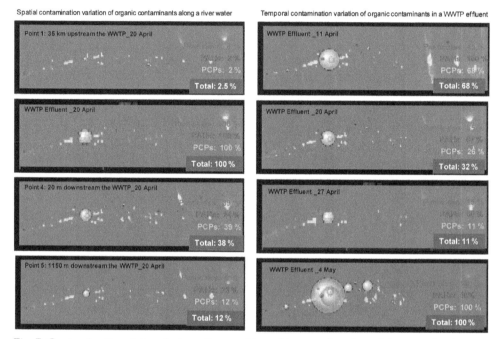

Fig. 7. Contamination status. Automatic searching of temporal and spatial contamination variation of organic contaminants. Reprinted with permission from Analytical Chemistry, (Gómez et al., 2011). Copyright 2011 American Chemical Society.

The most frequently detected contaminants and the contaminants detected at higher concentrations were the personal care products (musk fragrances galaxolide and tonalide). The pesticides and PAHs were detected at much lower concentration.

Halogenated compounds were successfully detected (Hashimoto et al., 2011) from several kinds of environmental samples by using a GC×GC chromatograph coupled with a tandem mass spectrometer (GC×GC–MS/MS). The global and selective detection of halogenated compounds was achieved by neutral loss scans of chlorine, bromine and/or fluorine using an MS/MS which was especially effective for compounds with more than two halogen substituents (Figure 8).

Screening and identification of pollutants was performed with GC×GC–HRTOFMS under the same conditions as those used for GC×GC–MS/MS. A lot of dioxins and PCBs congeners and many other compounds were identified in fly ash extract without any cleanup process and in sediment samples. In the future, the authors expect to achieve the complete global detection of any compound in one measurement of a crude sample simply with a GC×GC–HRTOFMS if it becomes possible to extract the desired information from the GC×GC-HR TOFMS data.

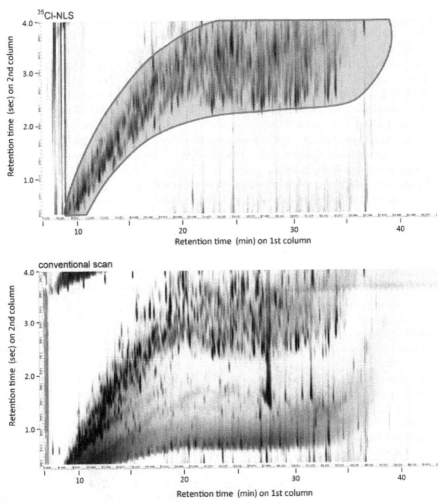

Fig. 8. Two-dimensional TICs of fly ash extract (NIES CRM17) measured with a ³⁵Cl-NLS (upper) and a conventional scan (lower) obtained with the GC×GC–MS/MS. The red translucent shape in the upper chromatogram shows the area where organohalogens are expected to appear. (For interpretation of the references to color in this figure legend, the reader is referred to the web version of the article.). Reprinted from Journal of Chromatography, A, (Hashimoto et al., 2011). Copyright (2011), with permission from Elsevier.

10. GC×GC instrumentation and optimization of operating conditions

Publications and reviews were dedicated to specific aspect of GC×GC. Major innovations in GC×GC modulator development were recently reviewed (Edwards et al., 2011). Cryogenic modulators remain very popular because of their ability to produce very small peak widths at half height and minimize breakthrough. Their commercial availability from several suppliers, has also contributed to their popularity. The use of valve-based modulators is increasing because of their less operating cost and easier maintenance than cryogenic modulator. Nevertheless, their coupling to MS remains problematic due to the large carrier gas flows in the second column and these modulators are not able to produce peaks of the same quality. Thermal modulation with the use of thermoelectric cooling could be a promising alternative if temperatures can be lowered enough to trap VOCs. Tranchida *et al.* (Tranchida et al., 2011a) have also published a review focused on the history (1991–2010) and present trends and future prospects for GC×GC modulation. Authors provided detailed descriptions and discussed the advantages and the disadvantages of the most significant thermal and pneumatic modulators. The authors have concluded that if at the moment, dual stage liquid N_2 systems can still be considered as the most effective modulators, in the next 10 years, the popularity of pneumatic modulators will gradually increase. The authors have included the description of their simple flow modulator, a seven port metallic disc published in 2011 (Tranchida et al., 2011b). A rotary and diaphragm 6-port 2-position valves have been also evaluated as modulators for GC×GC (Lidster et al., 2011).

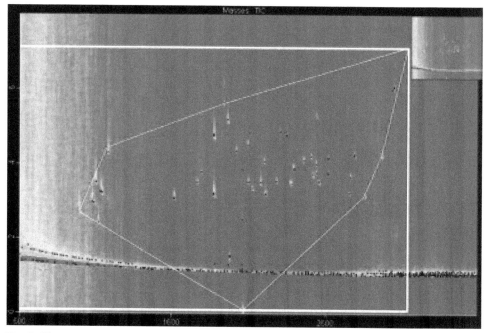

Fig. 9. Chromatogram illustrating the retention space used (white) and the space used calculated with Delaunay's triangulation algorithms (yellow) obtained on HP5-Mega225. Reprinted from Journal of Chromatography, A, (Semard et al., 2010). Copyright (2010), with permission from Elsevier.

In 2011, Panic *et al.* (Panic et al., 2011) developed a new consumable-free thermal modulator for GC×GC. The modulator was constructed from a trapping capillary, installed outside the GC oven, and coated inside with PDMS stationary phase. Dual-stage modulation was accomplished by resistively heating alternate segments of the trap with a custom-designed capacitive discharge power supply. The two unique inventions presented, flattening of the trap and selective removal of the stationary phase, have successfully eliminated the traditional drawbacks of resistively heated modulators.

The identification of compounds by using GC is based on peak retention times and mass spectra which generates uncertainty for the analyst for complex samples containing isomeric species. Retention index procedures were introduced to minimize misidentification of compounds in conventional chromatography. Various approaches to use of the retention index in GC×GC were reviewed and discussed (von Muhlen & Marriott, 2011).

A new method for the calculation of the percentage of separation space used was developed by Semard *et al.* (Semard et al., 2010) using Delaunay's triangulation algorithms (convex hull).

This approach was compared with an existing method and showed better precision and accuracy. It was successfully applied to the selection of the most convenient column set (HP5-Mega225) for the analysis of 49 target compounds including pesticides, HAPs, PCBs, PBDE *etc...* in wastewater. The diameter and length of the second column were optimized to improve the percentage of separation space used up to 40%.

Recently, Omais *et al.* (Omais et al., 2011) have shown that the general notion of orthogonality combining retention mechanisms independence and two dimensional space occupation must be decoupled. They have demonstrated that a non-orthogonal system can offer a good separation and a great space occupation. Moreover, orthogonality is intimately linked to the sample properties and cannot be considered as a sine qua none condition to achieve a good separation.

Table 1 summarizes the acronyms used in this review.

APEOs	AlkylPhenol EthOxylates
CTT	Toxaphene
CZC	Cryogenic Zone Compression
ECD, µECD	Electron Capture Detector, Micro Electron Capture Detector
EU	European Union
EPA	Environmental Protection Agency
FID	Flame Ionization Detector
GC	Gas Chromatography
GC×GC	Comprehensive two-dimensional gas chromatography
HS	HeadSpace
HRMS	High Resolution Mass Spectrometry
HRTOF	High resolution time-of-flight
ID	Isotope Dilution
ILs	Ionic Liquids
LLE	Liquid Liquid Extraction
LOD	Limit of detection
MS	Mass Spectrometer
MS/MS	Tandem mass spectrometer
NIST	National Institute of Standards and Technology
NPs	NonylPhenols

OCPs	OrganoChlorinated Pesticides
OPs	OctylPhenols
PAHs	Polycyclic Aromatic Hydrocarbons
Br-PAHs	Brominated Polycyclic aromatic hydrocarbon
Cl-PAHs	Chlorinated polycyclic aromatic hydrocarbon
PBDEs	PolyBrominated DiphenylEthers
PCBs	PolyChlorinated Biphenyls
PCDDs	PolyChlorinated Dibenzo-p-Dioxins
PCDFs	PolyChlorinated DibenzoFurans
PCNs	PolyChlorinated Naphthalenes
PCTs	PolyChlorinated Terphenyls
PDMS	PolyDiMethylSiloxane
POPs	Persistent Organic Pollutants
qMS	Quadrupole Mass Spectrometer
QSRR	Quantitative Structure–Retention Relationship
SBSE	Stir Bar Sorptive Extraction
SIM	Selected Ion Monitoring
SPE	Solid Phase Extraction
SPME	Solid Phase MicroExtraction
2,3,7,8-TCDD	2,3,7,8-tetrachlorodibenzo-p-dioxin
TD	Thermal Desorption
TEQ/kg	Toxic Equivalent Quantity/kg
TOF	Time-Of-Flight
US EPA	United States Environmental Protection Agency
VOCs	Volatile Organic Compounds
WFD	Water Framework Directive
WHO	World Health Organization
WWTPs	WasteWater Treatment Plants

Table 1. List of acronyms

11. Conclusion

This work reviews about 40 publications over the period 2009-2011 dealing with GC×GC and more especially on environmental applications. This technique coupled to mass spectrometry is an excellent choice for analyzing complex environmental samples and is suitable for multiresidue and non target analyses. As the number of environmental regulated pollutants increases, GC×GC appears ideal for the development of global detection method which can be used to screen a large number of compounds simultaneously. This kind of method could reduce cost and time necessary for their detection and quantification. GC×GC provides the analytical chemist with a new tool for a better separations of organohalogen congeners and, potentially, for more accurate human and environmental exposure data for risk assessments. Moreover, the authors highlighted the improved resolution and sensitivity offered by GC×GC over conventional one dimensional GC. The high selectivity of GC×GC-TOFMS has also facilitated the development of a wide range of analytical methods with minimal sample preparation and allowed the screening of emerging contaminants. Table 2 summarizes the main technical characteristics of methods applied to the various classes of pollutants reviewed in this chapter.

Compounds	Samples	Extraction	Material	Set of columns	references
PCBs and chlorobenzenes	Soil, sediment and sludge	ASE	GC×GC-µECD	DB1×Rtx-PCB	Muscalu et al., 2011
PCDDs and PCDFs	soil and sediment samples from South Africa	ASE	GC×GC-TOFMS	Rtx-Dioxin 2ₓRtx-PCB Rxi-5 Sil MSₓRtx-200 Rxi-XLBₓRtx-200	de Vos et al., 2011a
PCDDs, PCDFs and four dioxin-like non-ortho substituted PCBs	samples from a hazardous waste treatment facility	Soxhlet	GC×GC-TOFMS	Rtx-5Silms×Rtx-PCB Rtx5ₓRtx-200 Rtx-XLBₓRtx-200 Rtx-Dioxin 2ₓRtx-PCB.	de Vos et al., 2011b
Chlorinated and brominated polycyclic aromatic hydrocarbon (Cl-/Br-PAHs)	Soil samples from Japan	Soxhlet	GC×GC-HRTOFMS	BPX5×BPX50 BPX5×LC-50HT	Ieda et al., 2011
Naphthalene, benzothiophene and their alkylated congeners	Marine sediment	Sonication	GC×GC-FID and GC×GC-TOFMS	not specified	Wardlaw et al., 2011
Petroleum hydrocarbons	Soil	ASE	GC×GC-FID and GC×GC-TOFMS	RTX-1×BPX50	Mao et al., 2001
Naphthalene, benzothiophene and their alkylated congeners	Marine sediment	Sonication	GC×GC-FID and GC×GC-TOFMS	not specified	Wardlaw et al., 2011
4-Nonylphenols	Wastewater	LLE	GC×GC-FID and GC×GC-TOFMS	DB-5 ms×Supelcowax 10	Eganhouse et al., 2009
Pesticides (propanil, fipronil, propiconazole, trifloxystrobin, permethrin, difenoconazole and azoxystrobin)	Sediments (Brazil)	Sonication	GC×GC-µECD	DB-5×DB-17ms HP-50×DB-1ms	Macedo da Silva et al., 2011
Pesticides	Tap water	SPME	GC×GC-qMS	SLB-5ms×SLB-IL59	Purcaro et al., 2011
OCPS, PAHs, PCBs and pharmaceuticals and personal care products	River water	SBSE	GC×GC-HRTOFMS	DB-5ₓBPX-50	Ochiai & Sasamoto
benzothiazoles, benzotriazoles and benzosulfonamides	River water and in wastewater from both the influent and the effluent of a WWTP	SPE	GC×GC–TOFMS	TRB-5MSₓTRB-50HT ZB-WAXₓTRB-5MS TRB-1701MSₓTRB-5MS	Jover et al., 2009
VOCs and semi-volatile compounds	Marine salts	HS-SPME	GC×GC-TOFMS	BPX5×BP20	Silva et al., 2010
PAHs, PBDEs, Pesticides, drugs, personal care products, anti-microbials, carcinogen and reprotoxic compounds	Urban wastewater	LLE	GC×GC-TOFMS	HP5ₓMegaWaxHT	Semard et al., 2008
Phthalates, PAHs and their heterocyclic analogs, PCBs, PBDEs, chloroalkyl phosphates, pesticides, and pesticide degradation products	Household dusts	ASE	GC×GC-TOFMS	Rxi5-MS×BPX50	Hilton et al., 2010
Pharmaceuticals, plasticizers, personal care products, acid herbicides, triazines, organophosphorous compounds, phenylureas, organochlorine biocides, PAHs, and benzothiazoles	River	SPE	GCₓGC-TOFMS	TRB-5MS×TRB-50HT	Matamoros et al., 2010a
Personal care products, PAHs, pesticides, cholesterol and its degradation products ...	River	SBSE	GCₓGC-TOFMS	Rtx-5×Rxi-17 Rtx-5×Rt-LC50	Gomez et al., 2011
Halogenated compounds including PCBs, PCDDs, PCDFs...	Fly ash extract, Sediment, soil	Soxhlet	GC×GC-MS/MS and GC×GC-HRTOFMS	InertCap 5MS×BPX50	Hashimoto et al., 2011

Table 2. characteristics of methods applied to the various classes of pollutants based on compounds, sample preparation techniques, GC×GC method andset of columns used.

12. References

Adahchour, M.; Beens, J.; Vreuls, R. J. J. & Brinkman, U. A. T. (2006). Recent developments in comprehensive two-dimensional gas chromatography (GC×GC). III. Applications for petrochemicals and organohalogens. *Trends in Analytical Chemistry*, Vol.25, No.7, (July-August 2006), pp. 726-741, ISSN 0165-9936

Adahchour, M.; Beens, J. & Brinkman, U. A .T. (2008). Recent developments in the application of comprehensive two-dimensional gas chromatography. *Journal of Chromatography, A*, Vol.1186, No.1-2, (April 2008), pp. 67-108, ISSN 0021-9673

Amador-Munoz, O.; Villalobos-Pietrini, R. ;Aragon-Pina, A.; Tran, T.C.; Morrison, P. & Marriott, P. J. (2008). Quantification of polycyclic aromatic hydrocarbons based on comprehensive two-dimensional gas chromatography-isotope dilution mass spectrometry. *Journal of Chromatography, A*, Vol.1201, No.2, (August 2008), pp. 161-168, ISSN 0165-9936

Arsene, C.; Vione, D.; Grinberg, N. & Olariu, R.I. (2011). GC×GC-MS hyphenated techniques for the analysis of volatile organic compounds in air. *Journal of Liquid Chromatography & Related Technologies*, Vol.34, No.13, (August 2011), pp. 1077-1111, ISSN 1082-6076

Ballesteros-Gomez, A. & Rubio, S. (2011). Recent Advances in Environmental Analysis. *Analytical Chemistry*, Vol.83, No.12, (April 2011), pp. 4579-4613, ISSN 0003-2700

Bester, K. (2009). Analysis of musk fragrances in environmental samples. *Journal of Chromatography, A*, Vol.1216, No.3, (January 2009), pp. 470-480, ISSN 0021-9673

Bordajandi, L. R.; Ramos, J. J.; Sanz, J.; Gonzalez, M. J. & Ramos, L. (2008). Comprehensive two-dimensional gas chromatography in the screening of persistent organohalogenated pollutants in environmental samples. *Journal of Chromatography, A*, Vol.1186, No.1-2, (april 2008), pp. 312-324, ISSN 0021-9673

Cortes, H. J.; Winniford, B.; Luong, J. & Pursch, M. Comprehensive two dimensional gas chromatography review.(2009). *Journal of Separation Science*, Vol.32, No.(5-6), (March 2009), pp 883-904, ISSN 1615-9306

D'Archivio,A. A.; Incani, A. & Ruggieri, F. (2011). Retention modelling of polychlorinated biphenyls in comprehensive two - dimensional gas chromatography. *Analytical and Bioanalytical Chemistry*, Vol.399, No. 2, (January 2011), pp. 903-913. ISSN 1618-2642

Dallüge, J.; Beens, J. & Brinkman U. A. T. (2003). Comprehensive two-dimensional gas chromatography: a powerful and versatile analytical tool. *Journal of Chromatography, A*, Vol.1000, No.1-2, (June 2003), pp. 69-108

de Vos, J.; Gorst-Allman, P. & Rohwer, E. (2011a). Establishing an alternative method for the quantitative analysis of polychlorinated dibenzo-p-dioxins and polychlorinated dibenzofurans by comprehensive two dimensional gas chromatography-time-of-flight mass spectrometry for developing countries. *Journal of Chromatography, A*, Vol.1218, No.21, (May 2011), pp. 3282-3290, ISSN 0021-9673

de Vos, J.; Dixon, R.; Vermeulen, G.; Gorst-Allman, P.; Cochran, J.; Rohwer, E. & Focant, J.-F. (2011b). Comprehensive two-dimensional gas chromatography time of flight mass spectrometry (GC × GC-TOFMS) for environmental forensic investigations in developing countries. *Chemosphere*, Vol.82, No.9, (February 2011), pp. 1230-1239, ISSN 0045-6535

Edwards, M.; Mostafa, A. & Gorecki, T. (2011). Modulation in comprehensive two – dimensional gas chromatography : 20 years of innovation. *Analytical and*

Bioanalytical Chemistry, DOI 10.1007/s00216-011-5100-6, no pp. yet given, ISSN 1618-2650.

Eganhouse, R. P.; Pontolillo, J.; Gaines, R. B.; Frysinger, Glenn S.; Gabriel, F. L. P.; Kohler, H.-P. E.; Giger, W. & Barber, L. B. (2009). Isomer-Specific Determination of 4-Nonylphenols Using Comprehensive Two-Dimensional Gas Chromatography/Time-of-Flight Mass Spectrometry. *Environmental Science & Technology*, Vol.43, No.24, (December 2009), pp. 9306-9313, ISSN 0013-936X

Feo, M. L.; Eljarrat, E.; Barcelo, D. & Barcelo, D. (2010). Determination of pyrethroid insecticides in environmental samples. *Trends in Analytical Chemistry*, Vol.29, No.7, (July-August 2010), pp. 692-705, ISSN 0165-9936

Giddings, J. C. (1984). Two-dimensional separations: concept and promise. *Analytical chemistry*, Vol.56, No.12, (October 1984), pp. 1258A-1260A, 1262A, 1264A, ISSN 0003-2700

Gomez, M. J.; Herrera, S.; Sole, D.; Garcia-Calvo, E.; Fernandez-Alba, A. R. (2011). Automatic Searching and Evaluation of Priority and Emerging Contaminants in Wastewater and River Water by Stir Bar Sorptive Extraction followed by Comprehensive Two-Dimensional Gas Chromatography-Time-of-Flight Mass Spectrometry. *Analytical Chemistry*, Vol.83, No.7, (April 2011), pp. 2638-2647

Hamilton, J. F. (2010). Using comprehensive two-dimensional gas chromatography to study the atmosphere. *Journal of Chromatographic Science*, Vol.48, No.4, (April 2010), pp. 274-282, ISSN 0021-9665

Hashimoto, S.; Takazawa, Y.; Fushimi, A.; Tanabe, K.; Shibata, Y.; Ieda, T.; Ochiai, N.; Kanda, H.; Ohura, T.; Tao, Q. & Reichenbach, S. E. (2011). Global and selective detection of organohalogens in environmental samples by comprehensive two-dimensional gas chromatography-tandem mass spectrometry and high-resolution time-of-flight mass spectrometry. *Journal of Chromatography, A*, Vol.1218, No.24, (June 2011), pp. 3799-3810, ISSN 0021-9673

Hilton, D. C.; Jones, R. S. & Sjoedin, A. (2010). A method for rapid, non-targeted screening for environmental contaminants in household dust. *Journal of Chromatography, A*, Vol.1217, No. 4, (October 2010), pp. 6851-6856, ISSN 0021-9673

Horii, Y.; Ok, G.; Ohura, T. & Kannan, K. (2008). Occurrence and Profiles of Chlorinated and Brominated Polycyclic Aromatic Hydrocarbons in Waste Incinerators. *Environmental Science & Technology*, Vol.42, No.6, (March 2008), pp. 1904-1909, ISSN 0013936X

Horii, Y.; Ohura, T.; Yamashita, N. & Kannan, K. (2009). Chlorinated Polycyclic Aromatic Hydrocarbons in Sediments from Industrial Areas in Japan and the United States. *Archives of Environmental Contamination and Toxicology*, Vol.57, No.4, (November 2009), pp. 651-660, ISSN 0090-4341

Ishaq, R.; Naf, C.; Zebuhr, Y.; Broman, D. & Jarnberg, U. (2003). PCBs, PCNs, PCDD/Fs, PAHs and Cl-PAHs in air and water particulate samples--patterns and variations. *Chemosphere*, Vol.50, No.9, (March 2003), pp. 1131-1150, ISSN 0045-6535

Ieda, T.; Ochiai, N.; Miyawaki, T.; Ohura, T. & Horii, Y. (2011). Environmental analysis of chlorinated and brominated polycyclic aromatic hydrocarbons by comprehensive two-dimensional gas chromatography coupled to high-resolution time-of-flight mass spectrometry. *Journal of Chromatography, A*, Vol.1218, No.21, (May 2011), pp. 3224-3232, ISSN 0021-9673

Jover, E.; Matamoros, V. & Bayona, J. M. (2009). Characterization of benzothiazoles, benzotriazoles and benzosulfonamides in aqueous matrixes by solid-phase extraction followed by comprehensive two-dimensional gas chromatography coupled to time-of-flight mass spectrometry. *Journal of Chromatography, A*, Vol.1216, No.18, pp. 4013-4019, ISSN 0021-9673

Lidster, R. T.; Hamilton, J. F. & Lewis, A. C. (2011). The application of two total transfer valve modulators for comprehensive two-dimensional gas chromatography of volatile organic compounds. *Journal of Separation Science*, Vol.34, No.7, (April 2011), pp. 812-821, ISSN 1615-9306

Liu, Z. & Phillips, J. B. (1991). Comprehensive two-dimensional gas chromatography using an on-column thermal modulator interface. *Journal of Chromatographic Science*, Vol.29, No.6, (June 1991), pp. 227-231, ISSN 0021-9665

Macedo da Silva, J.; Zini, C. A. & Caramao, E. B.(2011). Evaluation of comprehensive two-dimensional gas chromatography with micro-electron capture detection for the analysis of seven pesticides in sediment samples. *Journal of Chromatography, A*, Vol.1218, No.21, (May 2011), pp. 3166-3172, ISSN 0021-9673

Mao, D.; Lookman, R.; Van De Weghe, H.; Weltens, R.; Vanermen, G.; De Brucker, N. & Diels, L. (2009). Estimation of ecotoxicity of petroleum hydrocarbon mixtures in soil based on HPLC-GCXGC analysis. *Chemosphere*, Vol.77, No.11, (December 2009), pp. 1508-1513, ISSN0045-6535

Matamoros, V.; Jover, E. & Bayona, J.M. (2010a). Part-per-Trillion Determination of Pharmaceuticals, Pesticides, and Related Organic Contaminants in River Water by Solid-Phase Extraction Followed by Comprehensive Two-Dimensional Gas Chromatography Time-of-Flight Mass Spectrometry. *Analytical Chemistry*, Vol.82, No.2, (January 2010), pp. 699-706, ISSN0003-2700

Matamoros, V.; Jover, E. & Bayona, J. M. (2010b). Occurrence and fate of benzothiazoles and benzotriazoles in constructed wetlands. *Water Science and Technology*, Vol. 61, No. 1, (January 2010), pp. 191-198, ISSN 0273-1223

Muscalu, A. M.; Reiner, E. J.; Liss, S. N.; Chen, T.; Ladwig, G. & Morse, D. (2011). A routine accredited method for the analysis of polychlorinated biphenyls, organochlorine pesticides, chlorobenzenes and screening of other halogenated organics in soil, sediment and sludge by GCxGC-μECD. *Analytical and Bioanalytical Chemistry*, DOI 10.1007/s00216-011-5114-0, No pp. yet given., ISSN 1618-2642

Ochiai, N. & Sasamoto, K. (2011).Selectable one-dimensional or two-dimensional gas chromatography-olfactometry/mass spectrometry with preparative fraction collection for analysis of ultra-trace amounts of odor compounds.*Journal of Chromatography A*, Vol. 1218, No. 21, pp. 3180-3185, ISSN0021-9673

Omais, B.; Courtiade, M.; Charon, N.; Ponthus,J. & Thiebaut D. (2011). Considerations on Orthogonality Duality in Comprehensive Two-Dimensional Gas Chromatography, Analytical Chemistry, DOI 10.1021/ac201103e, No pp. yet given

Osemwengie, L. I. & Sovocool, G. W. (2011). Evaluation of comprehensive 2D gas chromatography -time-of-flight mass spectrometry for 209 chlorinated biphenyl congeners in two chromatographic runs. Chromatography Research International, doi:10.4061/2011/675920, 14 pp. ISSN 2090-3510

Pani, O. & Gorecki, T. (2006). Comprehensive two-dimensional gas chromatography (GC×GC) in environmental analysis and monitoring. Analytical and Bioanalytical Chemistry, Vol.386, No.4, (October 2006), pp. 1013-1023, ISSN 1618-2642

Panic, O.; Gorecki, T.; McNeish, C.; Goldstein, A. H.; Williams, B. J.; Worton, D. R.; Hering, S. V. & Kreisberg, N. M. (2011). Development of a new consumable-free thermal modulator for comprehensive two-dimensional gas chromatography. Journal of Chromatography, A, Vol.1218, No.20, (May 2011), pp. 3070-3079, ISSN 0021-9673.

Patterson, D. G. Jr.; Welch, S. M.; Turner, W. E.; Sjodin, A. & Focant, J.-F. (2011). Cryogenic zone compression for the measurement of dioxins in human serum by isotope dilution at the attogram level using modulated gas chromatographycoupled to high resolution magnetic sector mass spectrometry. Journal of Chromatography, A, Vol.1218, No.21, (May 2011), pp. 3274-3281, ISSN 0021-9673

Purcaro, G.; Tranchida, P. Q.; Conte, L.; Obiedzinska, A.; Dugo, P.; Dugo, G. & Mondello, L. (2011). Performance evaluation of a rapid-scanning quadrupole mass spectrometer in the comprehensive two - dimensional gas chromatography analysis of pesticides in water. Journal of Separation Science, Vol.34, No.18, (September 2011), pp. 2411-2417, ISSN 1615-9306

Ramos, J. J.; Pena-Abaurrea, M. & Ramos L. (2009). Environmental Analysis, In: Comprehensive Analytical Chemistry, Ramos Lourdes (Ed.), pp. 243-280, Elsevier Science, ISBN 0444532374, Amsterdam, The Netherlands

Schoenmakers, P.; Marriott, P. & Beens, J. (2003). Nomenclature and conventions in comprehensive multidimensional chromatography. LC-GC Europe, Vol.16, No.6, (June 2003), pp. 335-336 and 338-339, ISSN 1471-6577

Semard, G.; Bruchet, A.; Cardinael P. & Bouillon, J.-P. (2008a). Use of comprehensive two-dimensional gas chromatography for the broad screening of hazardous contaminants in urban wastewaters. Water Science and Technology, Vol.57, No.12, (June 2008), pp. 1983-1989, ISSN 0273-1223

Semard, G.; Bruchet, A.; Cardinael, P. & Peulon-Agasse, V. (2008b). Study of the significance of comprehensive two-dimensional gas chromatography for analysis of complex environmental matrices. Spectra Analyse, Vol.37, No.261, (April-May 2008), pp. 29-33, ISSN 1635-947X

Semard, G.; Adahchour, M. & Focant, J.-F. (2009). Basic Instrumentation for GC×GC, In: Comprehensive Analytical Chemistry, Ramos Lourdes (Ed.), pp. 243-280, Elsevier Science, ISBN 0444532374, Amsterdam, The Netherlands

Semard, G.; Peulon-Agasse, V.; Bruchet, A. ; Bouillon, J.-P. & Cardinael, P. (2010). Convex hull: A new method to determine the separation space used and to optimize operating conditions for comprehensive two-dimensional gas chromatography. Journal of Chromatography, A, Vol.1217, No.33, (August 2010), pp. 5449-5454, ISSN 0021-9673

Silva, I.; Rocha, S. M.; Coimbra, M. A. & Marriott, P. J. (2010). Headspace solid-phase microextraction combined with comprehensive two-dimensional gas chromatography time-of-flight mass spectrometry for the determination of volatile compounds from marine salt. Journal of Chromatography, A, Vol.1217, No.34, (August 2010), pp. 5511-5521, ISSN 0021-9673

Skoczynska, E.; Korytar, P & de Boer, J. (2008). Maximizing chromatographic information from environmental extracts by GCxGC-ToF-MS. *Environmental Science & Technology*, Vol.42, No.17, (September 2008), pp. 6611–6618, ISSN 0013-936X

Tranchida, P. Q.; Purcaro, G.; Dugo, P. & Mondello, L. (2011a). Modulators for comprehensive two-dimensional gas chromatography. *Trends in Analytical Chemistry*, Vol.30, No.9, (October 2011), pp.1437-1461, ISSN 0165-9936

Tranchida, P. Q.; Purcaro, G.; Visco, A.; Conte, L.; Dugo, P.; Dawes, P. & Mondello, L. (2011). A flexible loop-type flow modulator for comprehensive two-dimensional gas chromatography. *Journal of Chromatography, A*, Vol.1218, No.21, pp. 3140-3145, ISSN 0021-9673

Vallejo, A.; Olivares, M.; Fernandez, L. A.; Etxebarria, N.; Arrasate, S.; Anakabe, E.; Usobiaga, & A.Zuloaga, O. (2011). Optimization of comprehensive two dimensional gas chromatography-flame ionization detection-quadrupole masss pectrometry for the separation of octyl- and nonylphenol isomers. *Journal of Chromatography, A*, Vol. 1218, No.20, pp. 3064-3069, ISSN0021-9673

Wang, Y.; Chen, Q.; Norwood, D. L. & McCaffrey, J. (2010). Recent development in the application of comprehensive two-dimensional gas chromatograph. *Journal of Liquid Chromatography & Related Technologies*, Vol.33, No.9-12, (July 2010), pp. 1082-1115, ISSN 1082-6076

Wardlaw, G. D.; Nelson, R. K.; Reddy, C. M. & Valentine, D. L. (2011). Biodegradation preference for isomers of alkylated naphthalenes and benzothiophenes in marine sediment contaminated with crude oil. *Organic Geochemistry*, Vol.42, No.6, (July 2011), pp. 630-639, ISSN0146-6380

von Muhlen, C. & Marriott, P. J. (2011). Retention indices in comprehensive two - dimensional gas chromatography. *Analytical and Bioanalytical Chemistry*, DOI: 10.1007/s00216-011-5247-1, No pp. yet given. ISSN 1618-2642

Zapadlo, M.; Krupcik, J.; Majek, P.; Armstrong, D. W. & Sandra, P. (2010). Use of a polar ionic liquid as second column for the comprehensive two - dimensional GC separation of PCBs. *Journal of Chromatography, A*, Vol.1217, No.37, (September 2010), pp. 5859-5867, ISSN0021-9673

Zapadlo, M.; Krupcik, J.; Kovalczuk, T.; Majek, P.; Spanik, I.; Armstrong, D. W. & Sandra, P. (2011). Enhanced comprehensive two – dimensional gas chromatographic resolutionof polychlorinatedbiphenyls on a non-polar polysiloxane and an ionic liquid column series. *Journal of Chromatography, A*, Vol.1218, No.5, (February 2011), pp. 746-751, ISSN 0021-9673

Permissions

The contributors of this book come from diverse backgrounds, making this book a truly international effort. This book will bring forth new frontiers with its revolutionizing research information and detailed analysis of the nascent developments around the world.

We would like to thank Professor Dr. Mustafa Ali Mohd, for lending his expertise to make the book truly unique. He has played a crucial role in the development of this book. Without his invaluable contribution this book wouldn't have been possible. He has made vital efforts to compile up to date information on the varied aspects of this subject to make this book a valuable addition to the collection of many professionals and students.

This book was conceptualized with the vision of imparting up-to-date information and advanced data in this field. To ensure the same, a matchless editorial board was set up. Every individual on the board went through rigorous rounds of assessment to prove their worth. After which they invested a large part of their time researching and compiling the most relevant data for our readers. Conferences and sessions were held from time to time between the editorial board and the contributing authors to present the data in the most comprehensible form. The editorial team has worked tirelessly to provide valuable and valid information to help people across the globe.

Every chapter published in this book has been scrutinized by our experts. Their significance has been extensively debated. The topics covered herein carry significant findings which will fuel the growth of the discipline. They may even be implemented as practical applications or may be referred to as a beginning point for another development. Chapters in this book were first published by InTech; hereby published with permission under the Creative Commons Attribution License or equivalent.

The editorial board has been involved in producing this book since its inception. They have spent rigorous hours researching and exploring the diverse topics which have resulted in the successful publishing of this book. They have passed on their knowledge of decades through this book. To expedite this challenging task, the publisher supported the team at every step. A small team of assistant editors was also appointed to further simplify the editing procedure and attain best results for the readers.

Our editorial team has been hand-picked from every corner of the world. Their multi-ethnicity adds dynamic inputs to the discussions which result in innovative outcomes. These outcomes are then further discussed with the researchers and contributors who give their valuable feedback and opinion regarding the same. The feedback is then collaborated with the researches and they are edited in a comprehensive manner to aid the understanding of the subject.

Apart from the editorial board, the designing team has also invested a significant amount of their time in understanding the subject and creating the most relevant covers. They scrutinized every image to scout for the most suitable representation of the subject and create an appropriate cover for the book.

The publishing team has been involved in this book since its early stages. They were actively engaged in every process, be it collecting the data, connecting with the contributors or procuring relevant information. The team has been an ardent support to the editorial, designing and production team. Their endless efforts to recruit the best for this project, has resulted in the accomplishment of this book. They are a veteran in the field of academics and their pool of knowledge is as vast as their experience in printing. Their expertise and guidance has proved useful at every step. Their uncompromising quality standards have made this book an exceptional effort. Their encouragement from time to time has been an inspiration for everyone.

The publisher and the editorial board hope that this book will prove to be a valuable piece of knowledge for researchers, students, practitioners and scholars across the globe.

List of Contributors

Rong Lu, Takayuki Honda and Tetsuo Miyakoshi
Department of Applied Chemistry, School of Science and Technology, Meiji University, 1-1-1 Higashi-mita, Tama-ku, Kawasaki-shi, Japan

Cristina Macci, Serena Doni, Eleonora Peruzzi, Brunello Ceccanti and Grazia Masciandaro
Consiglio Nazionale delle Ricerche (CNR), Istituto per lo Studio degli Ecosistemi (ISE), Pisa, Italy

Jaya Shankar Tumuluru and Christopher T. Wright
Biofuels and Renewable Energies Technologies, Idaho National Laboratory, Idaho Falls, USA

Shahab Sokhansanj
Bioenergy Resource and Engineering Systems Group, Environmental Sciences Division, Oak Ridge National Laboratory, Oak Ridge, USA

Timothy Kremer
Chemical and Biomolecular Engineering Department, The Ohio State University, Columbus, USA

Shigeki Wada and Takeo Hama
Shimoda Marine Research Center, University of Tsukuba, Life and Environmental Sciences, University of Tsukuba, Japan

Pranav Kumar Gutch
Defence Research & Development, Establishment Jhansi Road, Gwalior, India

Huang Zeng, Fenglou Zou, Eric Lehne, Julian Y. Zuo and Dan Zhang
Schlumberger DBR Technology Center, Edmonton, AB, Canada

Peter Kusch
Bonn-Rhine-Sieg University of Applied Sciences, Department of Applied Natural Sciences, Rheinbach, Germany

Solomon Tesfalidet
Department of Chemistry, Umeå University, Sweden

Kasylda Milczewska and Adam Voelkel
Poznan University of Technology, Institute of Chemical Technology and Engineering, Poznan, Poland

Douglas Lane
Environment Canada, Canada

Ji Yi Lee
Chosun University, Republic of Korea

Cardinaël Pascal and Peulon-Agasse Valérie
SMS, Université de Rouen, France

Bruchet Auguste
CIRSEE (Centre International de Recherche Sur l'Eau et l'Environnement), France

9 781632 381293